NUCLEAR WAR AND ENVIRONMENTAL CATASTROPHE

NUCLEAR WAR AND ENVIRONMENTAL CATASTROPHE

NOAM CHOMSKY AND LARAY POLK

Seven Stories Press
NEW YORK • EMERYVILLE

Seven Stories Press
140 Watts Street
New York, NY 10013
www.sevenstories.com

College professors and middle and high school teachers may order free examination copies of Seven Stories Press titles. To order visit http://www.sevenstories.com/textbook or send a fax on school letterhead to (212) 226-1411.

Book design by Elizabeth DeLong

Library of Congress Cataloging-in-Publication Data
Chomsky, Noam.
  Nuclear war and environmental catastrophe / Noam Chomsky and Laray Polk. -- 1st ed.
    p. cm.
  Includes bibliographical references and index.
  ISBN 978-1-60980-454-1 (pbk.)
 1. Nuclear warfare--Environmental aspects. 2. Environmental disasters. I. Polk, Laray. II. Title.
  QH545.N83C56 2013
  363.325'5--dc23

                          2012046137

Printed in the United States

9 8 7 6 5 4 3 2 1

Contents

Preface

If humans choose to work to minimize the existential threats of our time, perhaps the most improbable aspect of remedy is that we will accept modalities based on collaboration and creative adaptation, rather than perpetual combat and domination.[1] It is a stark fact that present and future economies are predicated on a finite energy resource: carbon-based fuels.[2] Consensual science on climate change presents another fact: we may only have a few years to make adjustments in the collective carbon load before we are faced with irreversible consequences. As Christian Parenti in *Tropic of Chaos* perceptively and correctly points out:

> "[E]ven if all greenhouse gas emissions stopped immediately—that is, if the world economy collapsed today, and not a single light bulb was switched on nor a single gasoline-powered motor started ever again—there is already enough carbon dioxide in the atmosphere to cause *significant* warming and disruptive climate change, and with that considerably more poverty, violence, social dislocation, forced migration, and political upheaval. Thus

we must find humane and just means of adaptation, or we face barbaric prospects."[3]

Seen in this light, to live collaboratively and creatively is less a radical proposal than a pragmatic one, if we, future generations, and the biosphere are to survive nuclear war and environmental catastrophe.

Laray Polk
Dallas, Texas
September 2012

Abbreviations

ACHRE	Advisory Committee on Human Radiation Experiments
AEC	Atomic Energy Commission
ALEC	American Legislative Exchange Council
API	American Petroleum Institute
ARPA-E	Advanced Research Projects Agency-Energy
BIOT	British Indian Ocean Territory
BLEEX	Berkeley Lower Extremity Exoskeleton
BP	British Petroleum
CDB	China Development Bank
CIA	Central Intelligence Agency
CND	Campaign for Nuclear Disarmament
COP	Conference of the Parties to the UNFCCC
CTBT	Comprehensive Test Ban Treaty
CW	chemical weapons
DARPA	Defense Advanced Research Projects Agency

DEFCON	defense readiness condition
DOD	Department of Defense
DOE	Department of Energy
DU	depleted uranium
EPA	Environmental Protection Agency
GE	General Electric
HEU	highly enriched uranium
IAEA	International Atomic Energy Agency
IBM	International Business Machines
ISN	Institute for Soldier Nanotechnologies
IT	Information Technology
LEU	low-enriched uranium
MAD	mutually assured destruction
MIT	Massachusetts Institute of Technology
NAM	Non-Aligned Movement
NATO	North Atlantic Treaty Organization
NAVSTAR GPS	navigation system for timing and ranging, Global Positioning System
NEPA	National Environmental Policy Act
NIH	National Institutes of Health
NNI	National Nanotechnology Initiative
NPT	Non-Proliferation Treaty
NSC	National Security Council
NSF	National Science Foundation
NSG	Nuclear Suppliers Group

NWFZ	nuclear-weapon-free zone
OPEC	Organization of the Petroleum Exporting Countries
OSRD	Office of Scientific Research and Development
PNE	peaceful nuclear explosion
POW	prisoner of war
PTBT	Partial Test Ban Treaty
R&D	research and development
RADAR	radio detection and ranging
SDS	Students for a Democratic Society
START	Strategic Arms Reduction Treaty
TRIPS	Trade-Related Aspects of Intellectual Property Rights
UN	United Nations
UNFCCC	UN Framework on Convention on Climate Change
WgU	weapon-grade uranium
WTO	World Trade Organization

1.

Environmental Catastrophe

LARAY POLK: When we began this conversation in 2010, our starting point was a statement you had recently made in the press: "There are two problems for our species' survival—nuclear war and environmental catastrophe." What is meant by "environmental catastrophe"?

NOAM CHOMSKY: Actually, quite a lot of things. The major one is anthropogenic global warming—human contribution to global warming, greenhouse gases, others—but that's only a part of it. There are other sources of what's called pollution—the destruction of the environment—that are quite serious: erosion, the elimination of agricultural land, and turning agricultural land into biofuel, which has had a severe effect on hunger. It's not just an environmental problem; it's a human problem. Building dams and cutting down the Amazon forests has ecological consequences—there are thousands of things and the problems are getting a lot worse.

For one reason, because of the role of the United States. I mean, nobody's got a wonderful role in this, but as long as the United States is dragging down the entire world, which is what it's doing now, nothing significant is going to happen on these issues. The US has to at least be seriously taking part and should be well in the lead. It's kind of ironic; if you look at this hemisphere, the country that is well in the lead in trying to do something serious about the environment is the poorest country in South America, Bolivia. They recently passed laws granting rights to nature.[1] It comes out of the indigenous traditions, largely—the indigenous majority, they've got the government advocating on their behalf. Sophisticated Westerners can laugh at that, but Bolivia is going to have the last laugh.

Anyway, they're doing something. In the global system, they're in the lead, along with indigenous communities in Ecuador. Then there's the richest country—not only in the hemisphere, but in world history—the richest, most powerful country, which is not only doing nothing, but is going backward. Congress is now dismantling some of the legislation and institutions put into operation by our last liberal president, Richard Nixon, which is an indication of where we are.[2] In addition, there's a great enthusiasm about tapping new sources of fossil fuels and doing it in ways which are extremely environmentally destructive: water and other resources are destroyed through fracking and deep-sea drilling.[3] Anywhere you can find anything that you can use to destroy the environment, they're going after it with great enthusiasm. It's like issuing a death sentence on the species.

And what makes it worse is that a lot of it is being done out of principle—that it's not problematic, that it's what we ought to be doing. In a sense, the same is true of nuclear weapons. They're justified on the grounds that we need them for defense—we don't need them for defense—but the argument for moving forward toward disaster is a conscious, explicit argument that is widely believed. With regard to the environment and the United States, there is also quite a substantial propaganda campaign, funded by the major business organizations, which are quite frank about it. The US Chamber of Commerce and others are trying to convince people that it's not our problem, or that it's not even real.[4]

If you look at the latest Republican primary campaign, virtually every participant simply denies climate change. One candidate, Jon Huntsman, said he thinks it is real, but he was so far out of the running, it didn't matter.[5] Michele Bachmann said something to the effect, "Well, it could be real, but if it is, it's God's punishment for allowing gay marriage."[6] Whatever the world thinks, they can't do much if this is going on in the United States.

In Congress, among the latest cohort of Republican House representatives from 2010, almost all are global-warming deniers and are acting to cut back legislation to block anything meaningful, and to roll back the little that exists. I mean, it's surreal. If someone were watching this from Mars, they wouldn't believe what was happening on Earth.

Hugo Chávez gave a speech at the United Nations at one of the major General Assembly meetings, and, of course, the press was full of ridicule and absurdities and so on. They didn't mention the talk he gave. You can find the talk, I'm sure, on the Internet, in which he said that producers and consumers are going to have to get together and find ways to reduce reliance on hydrocarbons and fossil fuels.[7] Of course, Venezuela is a major oil producer. In fact, practically the whole economy depends upon it; they're a lot more reliant on oil than Texas is. So it can be done. We don't have to be lunatics who are willing to sacrifice our grandchildren so that we can have a little more profit.

Actually, the whole Texas system is interesting. I'm sure you know the history, but back around 1958, the Eisenhower administration introduced an arrangement whereby the United States would rely on Texas oil: exhaust our domestic oil resources instead of using much cheaper and more accessible Saudi oil resources, for the benefit of Texas oil producers.[8] And I think for the next fourteen years, the country relied primarily on Texas oil. Meaning, exhaust domestic resources and later on dig holes in the ground and pour oil back into them for strategic reserve. This was pretty sharply criticized even from a straight security point of view. An MIT faculty member, M. A. Adelman, an economist who is an oil specialist, testified before Congress on this, but it didn't matter. Profits for Texas oil producers overwhelm even elementary security considerations like reliance on foreign oil.

That's what it means to have a country that's business-run, nothing else matters. It is the same reason we can't have a health system like every other industrial country. The people who matter, the financial institutions, won't allow it so it's off the agenda.

The Koch brothers give large amounts of money to universities. In exchange they get a hand in choosing faculty.[9] How corrosive is this practice?

Such practices would be extremely harmful, virtually by definition. If universities (journals, researchers) are to serve their public function in a free and democratic society, the institutions and the faculty must be scrupulous in rejecting outside pressures, particularly from funders, whether these are state or private. Funding should be flatly rejected if it comes with conditions such as those you describe.

Nine out of twelve Republicans on the House Energy and Commerce Committee signed an Americans for Prosperity pledge to oppose regulation of greenhouse gases.[10] To what degree are campaign contributions and pledges like this one stifling the political process on environmental issues in the US? And to what degree does US energy policy impact other countries?

The US is the richest and most powerful state in the world, by a large measure. Its policies on anything impact others. Energy policies in particular have an enormous impact, also on future generations, for reasons too well known to spell out again in detail here. The Republican pledge is simply another indication of abandonment of any pretense of participating in the political system as an authentic parliamentary party, instead taking on the role of lockstep uniformity in service to wealth and power. Dismantling the (much too weak) regulatory apparatus is simply a way of informing future generations that we care nothing about their fate as long as we and those we serve profit now. There is no doubt that campaign contributions have a significant effect on party programs and eventual government decisions, hence undermining democracy, if we understand democracy to be a system in which government decisions reflect the will of the public, not the power of those who can purchase outcomes with substantial contributions.

What are the factors that have led to conservative think tanks, largely funded by industry interests such as the Kochs and ExxonMobil, being able to hold sway over consensual science in terms of public opinion?[11] Even if the science is hard to comprehend for most nonscientists, isn't it evident what industries that fund climate-change skepticism stand to gain?

It is entirely evident. Major industries and lobbying

organizations (US Chamber of Commerce, etc.) have been quite frank about their efforts to sway public opinion to question the overwhelming scientific consensus on the severe threat of anthropogenic global warming. There is no novelty in this. Industries that produce what they know to be lethal products (lead, tobacco, etc.) were able to use their wealth and power for long periods to continue their murderous activities unhampered.[12] The effects have been dire, and continue to be, but they are even more ominous in the case of the intensive efforts to undermine steps that might preserve the possibility of a decent life for future generations— with effects already evident, but only a foretaste of much worse that is all too likely to come.

Is the fossil-fuel industry monolithic?

The industry, like others, is dedicated to profit and market share, not to human welfare. But it is not immune to public pressures, and also recognizes that there is potential profit to be made from development of sustainable energy. The industry is mostly oligarchic, but not a monolith, and there are some conflicts within it. But in general, counting on the good will and altruism of participants in a semimarket system never makes sense, and in this case is virtually a commitment to disaster.

2.

Protest and Universities

LARAY POLK: What was the rationale for the protest when the Iranian students came to MIT in the 1970s?

NOAM CHOMSKY: There was a secret agreement made between MIT and the shah of Iran, which pretty much amounted to turning over the Nuclear Engineering Department to the shah. For some unspecified but probably large amount of money, MIT agreed to accept nuclear engineers from Iran to train in the United States; it could have become a nuclear weapons program. There was not much question about that. They called it nuclear energy. It was being pressed in Washington by Cheney, Rumsfeld, Kissinger, and Wolfowitz. They wanted Iran to develop nuclear facilities and they were allies at the time. That was pre-1979. Well, the story leaked, as these things tend to do. And what happened then was pretty interesting. The students got pretty upset about it, there was a lot of student protest, and then finally a referendum on campus. I think about 80 percent voted against it. Of course, that's not binding; that's student opinion.

But there was enough of a protest that they had to call a faculty meeting about it. Usually nobody goes to faculty meetings—too boring to go to—but at this meeting everybody showed up; it was huge. The proposal was presented by the administration, and then there was discussion. There were maybe five of us, I think, who stood up to oppose it, and it passed overwhelmingly.

How did you present your opposition?

MIT, first of all, shouldn't be taking support from states developing nuclear capabilities. And if the US government wants to do it, I'll protest that too, but it shouldn't be done here. That's not the task of a university to help other countries develop nuclear capabilities. They shouldn't do it here either, but certainly not for another country ruled by a brutal tyrant just because it's an ally. But it was a straightforward argument. Essentially, the students' argument.

It's quite interesting if you think about it sociologically. The faculty of today are the students of a few years ago, but the shift in institutional role completely changed their attitudes. So the students were outraged and the faculty thought it was fine.

Incidentally, there are worse cases than this not involving the universities, but the press won't report on them. Reagan and Bush were practically in love with

Saddam Hussein—after the US basically won the war for Iraq in the Iraq-Iran War, Bush wanted to increase aid to Iraq over a lot of objections from the Treasury Department and others, mostly on economic grounds, but he wanted it. In fact, in 1989 Bush invited Iraqi nuclear engineers to the United States for advanced training in nuclear weapons production.[1] Then, in 1990, he sent a high-level senatorial delegation to Iraq led by Bob Dole (later a presidential candidate), with Alan Simpson and other big shots, and their mission was to convey Bush's greetings to his friend Saddam and also to inform him that he should disregard criticisms that he hears in the American press. They promised Saddam that they would remove someone from the Voice of America who was being critical of Saddam. This was long after all of his worst atrocities— the Anfal massacre, the Halabja massacre—which the Reagan administration tried to cover up, said they were done by the Iranians, not by Saddam. All of that has been deep-sixed. You can find it in congressional hearings, but nobody will report it or comment on it.[2] That's even worse than what was happening with Iran under the shah. The record is not a very pretty one if you look at it, and it certainly is not one of trying to reduce proliferation.

How can we be sure that any government is going to remain stable? For example, when the Iranian students came, the shah was in power, nobody knew that the Islamic uprising was coming—

They don't care whether it's Islamic or not. Take, say, Pakistan, in the 1980s: Pakistan was under the rule of a dictator, the worst of their many dictators, Zia-ul-Haq, who was also pushing a radical Islamist agenda and was receiving extensive funding from Saudi Arabia. They were trying to Islamize society, that's when they were starting to set up these madrassas all over where the kids just study the Koran and radical Islam and so on. Saudi Arabia is the center of radical Islam, the most extreme fundamentalist state anywhere, and Reagan was supporting it.[3] They don't care about radical Islam.

Like al-Qaeda in Afghanistan, bin Laden before—

The US supported them. And in fact, they explained why. It had nothing to do with liberating Afghanistan. The head of the CIA mission in Islamabad, that's where the planning was going on, was frank about it. Basically he said, "We don't have any interest in liberating Afghanistan, what we want to do is kill Russians." And this is their chance. Brzezinski, to paraphrase, said things like, "This is great, it's paying the Russians back for Vietnam."[4] What were the Russians doing in Vietnam? Well, they were providing some limited support for resistance to US aggression, but that's a crime, so we've got to pay them back by killing Russians and if a million Afghans die, it's their problem.

Do you think it ever occurred to planners that al-Qaeda, Osama bin Laden, might come up with their own agenda?

Probably not. You can say the same about Hamas. Israel supported Hamas in the early days because they were a weapon against the secular PLO.[5] The US and Israel have quite consistently supported radical Islamic stands and it goes way back. Back in the early '60s—in fact, the '50s and the '60s—there was a major conflict in the Arab world between Nasser, who was the symbol of secular nationalism, and Saudi rulers, who were the guardians of extremist, radical Islam. Who did the US support? The Saudis, of course. They were afraid of secular nationalism.

The British too. There's a book by a British diplomatic historian—which will probably barely get reviewed in England—which is about Britain's support of radical Islam, and it was quite extreme.[6] Same reasons, secular nationalism is much more dangerous. Sometimes you catch a tiger by the tail and don't expect it. This is pretty much the same with Hezbollah. They developed in reaction to the US-backed Israeli invasion and occupation. That's the way the world works.

Sometimes it's called blowback. There are some analysts who argue that this is a self-defeating policy. But I'm not so convinced. I mean, the big mistake is supposed to have been installing the shah and overturning the parliamentary system, but it's hard to see where that was a mistake. For twenty-five years it kept Iran completely under control, kept the US in control over the energy system. Planning doesn't go much beyond that. If things work for twenty-five years, that's a success.

The Tehran Research Reactor requires highly enriched uranium fuel to operate; the same is true for MIT's reactor. The Department of Energy has told MIT it must convert to low-enriched fuel, but the head of engineering at the reactor said they're most likely not going to meet the deadline.[7]

I don't know anything about this.

It's contentious on two points. One, there's a reactor in a densely populated urban area. And two—

It's what they're using as fuel. So, you can do things with the high-energy fuel that you can't do with less enriched fuel?

According to a Boston Globe *story from 2009, the "MIT reactor could be converted quickly if it were willing to give up some performance." The same report states the reactor "brings in about $1.5 million a year from commercial work, which covers about 60 percent of the annual operating costs."[8]*

Where do they get that from?

I'm unsure of the specific entities, but mainly from producing radiotherapies.[9]

There hasn't been an inquiry at MIT into research, as far as I'm aware, since 1969. At that time, under the pressure of the student movement, there was a faculty/student inquiry. Actually I was on the committee, the Pounds Commission, which looked into MIT's finances and also into war-related activities on campus. It was pretty interesting. It turned out that nobody, even the administration, knew the financial details. It turned out that roughly half of the institute budget was running two classified military laboratories: Lincoln, and what's now the Draper Lab. The other half of the budget, I think, was approximately 90 percent funded by the Pentagon in those days.

The Pentagon, contrary to what people believe, is the greatest funder there is. They don't pay all that much attention to what you're doing; they just know that they're the way to funnel taxpayer money into the next stage of the economy. We did look into military work. It turns out there was no classified work on campus and no direct military-related work, but anything that's done is likely to have some military application. The only department that had any war-related work was the Political Science Department and it was being done under the rubric of a Peace Research Institute—straight out of Orwell—which had villas in Saigon where they were sending students for PhDs on counterinsurgency. And they were also running secret seminars in the Political Science Department on Vietnam strategy and so on. I found out when I was invited to take part in one.

Outside the Political Science Department, it was pretty clean. Now, if you take a look over the years, Pentagon funding has declined and funding from the NIH has increased. And I suspect almost everybody understands it. The reason is because the cutting edge of the economy is shifting to biology, away from an electronics-based economy, so you have to rip off the taxpayer in some different fashion. We don't have a free-market economy. Federal spending, government procurement, and other devices are huge components. Funding is also getting more corporatized. I suspect what is going on here is more corporate funding, and the corporate funding has a general cheapening effect.

Federal funding is long term, it's nonintrusive, and they just want things to be done. But if big corporations fund something, they're not interested in the future health of the economy. They want something for themselves. So it means that research becomes more short term, it becomes much more secret. Federal funding is completely open, but a corporation can impose secrecy; they can indicate you're not going to get refunded unless you keep it quiet. So it does impose secrecy. There are some famous cases that have come out; one big scandal even made the *Wall Street Journal*.

As long as it stays secret, they can do as they like.

Robert Barsky wrote that during the protests on the MIT campus in the 1960s, you held an extreme position even among the liberal faculty. Basically, you didn't believe shutting down the labs involved in military research was the solution, but rather, "universities with departments that work on bacterial warfare should do so openly."[10] What is to be gained by that approach?

These matters came to a head at the Pounds Commission. Its primary concern was the relation of MIT's academic/research program to the two military laboratories it administered, Lincoln and the Instrumentation Lab (now the Draper Lab). The commission split three ways. One group (call them "conservatives") favored keeping the labs on campus. A second ("liberals") favored separating the labs from MIT. A third ("radicals," I think consisting just of me and the one student representative) agreed with the conservatives, though for different reasons. If the labs were formally separated, nothing much would change in substance: joint seminars and other interactions would continue pretty much as before, but now with formally separate entities. What the labs are doing would disappear as a campus issue. But what they are doing is vastly more important than the appearance of a "clean campus," and their presence would be a regular focus for education and activism. The liberal view prevailed, and the outcome was much as anticipated—a step backward, I think, for the reasons mentioned.

Pentagon funding was a major device used by the government from the early postwar period to lay the basis for the high-tech economy of the future: computers, the Internet, microelectronics, satellites, etc.—the IT revolution generally. After several decades primarily in the public sector, the results were handed over to private enterprise for commercialization and profit. By the 1970s government funding was shifting from the Pentagon to the biology-related institutions: the NIH and others. The military was a natural funnel for an electronics-based economy. Fifty years ago the small start-ups spinning off from MIT were electronics firms, which, if successful, were bought up by Raytheon and other electronics giants. Today the small start-ups are in genetic engineering, biotechnology, etc., and the campus is surrounded by major installations of pharmaceutical firms and the like.[11] The same dynamics have been duplicated elsewhere.

The Pentagon itself gains little if anything from this, not even prestige. In fact, few even know how the system works. To illustrate, I once wrote an article about a speech to newspaper editors by Alan Greenspan—called "St. Alan" during his day in the sun, and heralded as one of the great economists of all time. He was hailing the marvels of our economy, based on entrepreneurial initiative and consumer choice, the usual oration. He made the mistake, however, of giving examples, each of them textbook illustrations of what I have just described: the role of the dynamic state sector of the economy during the hard part of research and development (along with government procurement and other devices of what amounts to a kind of industrial policy). Greenspan's illusions are the common picture.[12]

The system as a whole certainly merits critical examination, for one reason, because there is virtually no public input in crucial decision making. But I've never seen the force of the argument against employment at a university that is being publicly funded for research, development, and teaching, and it seems of little moment whether the funding technique happens to be via the Pentagon, the NIH, the Department of Energy, or some other formal mechanism.

In general, what matters is what work is being done, not how it's funded. Biological warfare is no more benign if it's funded by the NIH or by a private corporation. Universities are parasitic institutions. They don't (or shouldn't) be geared to production for the market. If they are to survive, they have to be funded somehow, and there are few options in existing society.

For what it's worth, while the MIT lab where I was working in the '60s was 100 percent funded by the armed services (as you can see from formal acknowledgments by publishers), it also happened to be one of the main centers of academic resistance against the Vietnam War, perhaps the main one; not protest, but active resistance.[13] And by the late '60s MIT probably had the most radical student president of any US campus, with plenty of student support and related activism, which had quite positive and long-lasting effects on campus life.[14]

What are alternative ways of viewing campuses?

A campus is primarily an educational institution. A crucial part of education is coming to understand the world in which we live, and what we can do to make it a better place. Any college, and particularly a research university like MIT, should also be a center of creative and independent thought and inquiry, along with critical evaluation of the directions that such inquiry should pursue, with cooperative participation of the general university community.[15] It should also bring in, as feasible, the outside community. My own courses on social and political issues—which I was teaching on my own time— were usually open to the public, sometimes at night for that reason, others too.

I don't suggest any of this as an "alternative," but rather as an ideal, approximated more or less, and a guideline for commitment and choice of action.

3.

Toxicity of War

LARAY POLK: We've spoken previously about Reagan's "Star Wars" program as something that developed as a palatable alternative to nuclear stockpiling. Isn't there also something to be said about the location of the Ronald Reagan Ballistic Missile Defense Test Site on Kwajalein Atoll, a heavily contaminated area as a result of US atomic testing?[1]

NOAM CHOMSKY: I suppose it reflects the prevailing conception that the "unpeople" of the world—to borrow the phrase of British diplomatic historian Mark Curtis—are dispensable.[2]

The level of contamination left from US atomic testing in the Marshall Islands is immensely troubling, but so too is what might be transpiring in Iraq and other areas of the Middle East due to the use of depleted uranium. There seems to be ample evidence that the use of DU in Iraq by the US is causing a catastrophic health crisis. Some have even referred to it as "low-grade nuclear warfare."³ Where do you stand on this issue?

The levels of birth defects, cancer, and other consequences of the US assault on Iraq are shocking. Whether the cause is DU remains uncertain—same in other areas. There are many sources of toxicity in warfare. The authors of the published studies have suggested that DU might be the cause, but report that they cannot be confident. To my knowledge, serious weapons specialists and nuclear scientists deeply concerned about these issues have reached no definite conclusions.

The people of Vietnam also suffer from an inordinate number of birth defects. In comparing that situation to the use of DU in Iraq, is it possible that the inability to reach definite conclusions about health and environmental issues is intentional?⁴ Are there political factors that get in the way of scientific research that has the potential to establish causation?

There is a valuable new study of the effects of Agent Orange on South Vietnamese by Fred Wilcox, *Scorched Earth: Legacies of Chemical Warfare in Vietnam*—a very serious work, beyond anything else I've seen. He had an earlier book on its effects on US soldiers: *Waiting for an Army to Die: The Tragedy of Agent Orange.*⁵ Since we last spoke there have been some investigations of the impact of US weaponry in the attack on Fallujah. One technical study found unusually high levels of enriched uranium, presumably from DU, along with other dangerous substances.[6] Another study, reported by Patrick Cockburn in the London *Independent* and in the *International Journal of Environmental Research and Public Health*, found that "Dramatic increases in infant mortality, cancer and leukaemia in the Iraqi city of Fallujah, which was bombarded by US Marines in 2004, exceed those reported by survivors of the atomic bombs that were dropped on Hiroshima and Nagasaki in 1945." The study, by Iraqi and British doctors, found "a four-fold increase in all cancers and a 12-fold increase in childhood cancer in under-14s. Infant mortality in the city is more than four times higher than in neighbouring Jordan and eight times higher than in Kuwait."[7]

That the marine attack on Fallujah in November 2004 (the second major assault) was a major war crime was evident at once even from the (generally supportive) US reporting. These new investigations surely merit widespread attention (they have received almost none) and serious inquiry, in fact war crimes trials, if that were imaginable. It is not: only the weak and defeated are subjected to such indignities.

There can hardly be any serious doubt that political factors interfere with scientific research in all such cases, massively in fact; and there are quite a few. The vicious US-supported Israeli invasion of Gaza in December 2008–January 2009 is another case that should be investigated.

The heroic Norwegian physicians Mads Gilbert and Erik Fosse, who worked under horrible conditions at the al-Shifa hospital in Gaza right through the worst days, reported effects of unknown lethal munitions that surely would receive extensive inquiry, bitter condemnation, and calls for punishment if the agents were enemy states.[8]

DU munitions in the US are produced by contractor-owned, contractor-operated facilities. Is this a way to deflect potential liability?[9]

In the case of Agent Orange, the US government claimed not to be aware that it contained dioxin, one of the most lethal known carcinogens. Wilcox provides evidence that the corporations providing the materials to the government were well aware of this, and chose not to remove the lethal components to save costs.[10] That Washington was unaware seems hardly credible, most likely an instance of what has sometimes been called "intentional ignorance." It should be remembered that when he escalated the attack on South Vietnam fifty years ago from support for a murderous client state to outright US aggression, President Kennedy authorized the use of chemical weapons to destroy ground cover and also food crops, a crime in itself, even apart from the dreadful scale and character of the consequences, with deformed fetuses to this day, several generations down the line in Saigon hospitals as a result of persistent genetic mutations.[11]

Somehow none of this exercises those who passionately proclaim their devotion to "right to life" even for the fertilized egg.

However, it is unlikely that the issue of liability will arise for the reason already mentioned. The powerful are self-immunized from even inquiry, let alone punishment for their crimes.[12]

The Radiation Protection Center in Baghdad has found a "clear radiation trail" from tanks hit by DU penetrators to their relocation to scrap sites.[13] What obligations to cleanup do the US and the UK have? How likely is it?

Obligations can be legal or moral. The US and UK invasion was a textbook example of the crime of aggression, "the supreme international crime differing only from other war crimes in that it contains within itself the accumulated evil of the whole," in the wording of the Nuremberg Tribunal, which sentenced Nazi war criminals to death for committing this crime. We should therefore have the honesty either to concede that the tribunal was judicial murder, hence our crime, or to recognize that George Bush, Tony Blair, and their accomplices should be subjected to the legal principles established at Nuremberg. Cleanup would be one important obligation on legal grounds, but a minor one in context. At the very least, the US and UK are obligated to provide massive reparations for their crimes against Iraq.

Judgment on moral grounds depends on what one's moral principles are. There is no doubt that cleanup—in fact, far more—would be regarded as a moral obligation if the crimes had been carried out by an enemy. Therefore it is an obligation for us if we are capable of accepting one of the most elementary of moral principles, found in every moral code worth consideration: the principle of universality, holding that we should apply to ourselves the standards we impose on others, if not more stringent ones.

How likely is it? Highly unlikely unless the dominant elites, particularly the educated classes, make an effort to rise to a level of civilization for which there is, unfortunately, no sign. In fact, even raising the issue arouses horror and often hysteria.

Does secrecy in matters of radioactive releases—intentional or unintentional—pose a danger equal to the materials themselves?[14]

Perhaps. But the greatest threat, I think, is the evasion and suppression of what is known, or could easily be known if there were any authentic concern for terrible crimes. Of course there is much anguish when someone else is guilty, but the crucial case, always, is when we ourselves are the perpetrators—clearly the most crucial case for us, on elementary moral grounds. Sometimes there is awareness, though ineffectual. Thomas Jefferson famously said that "I tremble for my country when I reflect that God is just, that his justice cannot sleep forever," referring to the crime of slavery. John Quincy Adams, the great grand strategist who was the intellectual author of Manifest Destiny, expressed very similar thoughts in reflecting on the "extermination" of the indigenous population with "merciless and perfidious cruelty . . . among the heinous sins of this nation, for which I believe God will one day bring [it] to judgement." Their concerns should resonate painfully to the present day. Those who preach most eloquently about their devotion to their Lord express only contempt for such thoughts; and they have plenty of company, needless to say. The US and its intellectual community are breaking no new ground, of course. They are following the course typical of systems of power, throughout history. We should, I think, take all of this as an indication of the great chasm that lies between the most advanced cultures and minimal standards of elementary decency, honesty, and moral integrity. Not a small problem, quite apart from the matters we are discussing.

4.

Nuclear Threats

LARAY POLK: What immediate tensions do you perceive that could lead to nuclear war? How close are we?

NOAM CHOMSKY: Actually, nuclear war has come unpleasantly close many times since 1945. There are literally dozens of occasions in which there was a significant threat of nuclear war. There was one time in 1962 when it was very close, and furthermore, it's not just the United States. India and Pakistan have come close to nuclear war several times, and the issues remain. Both India and Pakistan are expanding their nuclear arsenals with US support. There are serious possibilities involved with Iran—not Iranian nuclear weapons, but just attacking Iran—and other things can just go wrong. It's a very tense system, always has been. There are plenty of times when automated systems in the United States— and in Russia, it's probably worse—have warned of a nuclear attack which would set off an automatic response except that human intervention happened to take place in time, and sometimes in a matter of minutes.[1] That's playing with fire. That's a low-probability event, but with low-probability events over a long period, the probability is not low.

There is another possibility that, I think, is not to be dismissed: nuclear terror. Like a dirty bomb in New York City, let's say. It wouldn't take tremendous facility to do that. I know US intelligence or people like Graham Allison at Harvard who works on this, they regard it as very likely in the coming years—and who knows what kind of reaction there would be to that. So, I think there are plenty of possibilities. I think it is getting worse. Just like the proliferation problem is getting worse. Take a couple of cases: In September 2009, the Security Council did pass a resolution, S/RES/1887, which was interpreted here as a resolution against Iran. In part it was, but it also called on all states to join the Non-Proliferation Treaty. That's three states: India, Pakistan, and Israel. The Obama administration immediately informed India that this didn't apply to them; it informed Israel that it doesn't apply to them.[2]

If India expands its nuclear capacity, Pakistan almost has to; it can't compete with India with conventional forces. Not surprisingly, Pakistan developed its nuclear weapons with indirect US support. The Reagan administration pretended they didn't know anything about it, which of course they did.[3] India reacted to resolution 1887 by announcing that they could now produce nuclear weapons with the same yield as the superpowers.[4] A year before, the United States had signed a deal with India, which broke the pre-existing regime and enabled the US to provide them with nuclear technology—though they hadn't signed the Non-Proliferation Treaty. That's in violation of congressional legislation going back to India's first bomb, I suppose around 1974 or so. The United States kind of rammed it through the Nuclear Suppliers Group, and that opens a lot of doors. China reacted by sending nuclear technology to Pakistan. And though the claim is that the technology for India is for civilian use, that doesn't mean much even if India doesn't transfer that to nuclear weapons. It means they're free to transfer what they would have spent on civilian use to nuclear weapons.[5]

And then comes this announcement in 2009 that the International Atomic Energy Agency has been repeatedly trying to get Israel to open its facilities to inspection. The US along with Europe usually has been able to block it. And more significant is the effort in the international agencies to try to move toward a nuclear-weapon-free zone in the Middle East, which would be quite significant.[6] It wouldn't solve all the problems, but whatever threat Iran may be assumed to pose—and that's a very interesting question in itself, but let's suppose for the moment that there is a threat—it would certainly be mitigated and might be ended by a nuclear-weapon-free zone, but the US is blocking it every step of the way.[7]

Now that Iran's reactor at Bushehr is running, the current fear is that they're going to use the plutonium produced from the fuel cycle to make weapons. The questions raised about Iran's possible nuclear weapons program are similar to those asked of Israel—[8]

Since the 1960s. And in fact, the Nixon administration made an unwritten agreement with Israel that it wouldn't do anything to compel Israel, or even induce them, to drop what they call their ambiguity policy—not saying whether or not they have them.[9] That's now very alive because there's this regular five-year Non-Proliferation Review Conference. In 1995, under strong pressure from the Arab states, Egypt primarily, there was an agreement that they would move toward a nuclear-weapon-free zone and the Clinton administration signed on. It was reiterated in 2000. In 2005 the Bush administration just essentially undermined the whole meeting. They basically said, "Why do anything?"

It came up again in May 2010. Egypt is now speaking for the Non-Aligned Movement, 118 countries, they're this year's representative, and they pressed pretty hard for a move in that direction. The pressure was so strong that the United States accepted it in principle and claims to be committed to it, but Hillary Clinton said the time's "not ripe for establishing the zone."[10] And the administration just endorsed Israel's position, essentially saying, "Yes, but only after a comprehensive peace agreement in the region," which the US and Israel can delay indefinitely. So, that's basically saying, "it's fine, but it's never going to happen." And this is barely ever reported, so nobody knows about it. Just as almost nobody knows about Obama informing India and Israel that the resolutions don't apply to them. All of this just increases the risk of nuclear war.

It's more than that actually. You know, the threats against Iran are nontrivial and that, of course, induce them to move toward nuclear weapons as a deterrent. Obama in particular has strongly increased the offensive capacity that the US has on the island of Diego Garcia, which is a major military base they use for bombing the Middle East and Central Asia.[11] In December 2009, the navy dispatched a submarine tender for nuclear submarines in Diego Garcia. Presumably they were already there, but this is going to expand their capacity, and they certainly have the capacity to attack Iran with nuclear weapons. And he also sharply increased the development of deep-penetration bombs, a program that mostly languished under the Bush administration. As soon as Obama came in, he accelerated it, and it was quietly announced—but I think not reported here—that they put a couple of hundred of them in Diego Garcia. That's all aimed at Iran. Those are all pretty serious threats.[12]

Actually, the question of the Iranian threat is quite interesting. It's discussed as if that's the major issue of the current era. And not just in the United States, Britain too. This is "the year of Iran," Iran is the major threat, the major policy issue. It does raise the question: What's the Iranian threat? That's never seriously discussed, but there is an authoritative answer, which isn't reported. The authoritative answer was given by the Pentagon and intelligence in April 2010; they have an annual submission to Congress on the global security system, and of course discussed Iran.[13] They made it very clear that the threat is not military. They said Iran has very low military spending even by the standards of the region; their strategic doctrine is completely defensive, it's designed to deter an invasion long enough to allow diplomacy to begin to operate; they have very little capacity to deploy force abroad. They say if Iran were developing nuclear capability, which is not the same as weapons, it would be part of the deterrent strategy, which is what most strategic analysts take for granted, so there's no military threat. Nevertheless, they say it's the most significant threat in the world. What is it? Well, that's interesting. They're trying to extend their influence in neighboring countries; that's what's called destabilizing. So if we invade their neighbors and occupy them, that's stabilizing. Which is a standard assumption. It basically says, "Look, we own the world." And if anybody doesn't follow orders, they're aggressive.

In fact, that's going on with China right now. It's been a kind of a hassle, also hasn't been discussed much in the United States—but is discussed quite a lot in China, about control of the seas in China's vicinity. Their navy is expanding, and that's discussed here and described as a major threat. What they're trying to do is to be able to control the waters nearby China—the South China Sea, Yellow Sea, and so on—and that's described here as aggressive intent. The Pentagon just released a report on the dangers of China. Their military budget is increasing; it's now one-fifth what the US spends in Iraq and Afghanistan, which is of course a fraction of the military budget. Not long ago, the US was conducting naval exercises in the waters off China. China was protesting particularly over the plans to send an advanced nuclear-powered aircraft carrier, the USS *George Washington*, into those waters, which, according to China, has the capacity to hit Beijing with nuclear weapons—and they didn't like it. And the US formally responded by saying that China is being aggressive because they're interfering with freedom of the seas. Then, if you look at the strategic analysis literature, they describe it as a classic security dilemma where two sides are in a confrontation. Each regards what it's doing as essential to its security and regards the other side as threatening its security, and we're supposed to take the threat seriously. So if China is trying to control waters off its coast, that's aggression and it's harming our security. That's a classic security dilemma. You could just imagine if China were carrying out naval exercises in the Caribbean—in fact, in the mid-Pacific—it would be considered intolerable. That's very much like Iran. The basic assumption is "We own the world," and any exercise of sovereignty within our domains, which is most of the world, is aggression.

Is there any type of nuclear racism involved in these issues?

I think it would be the same if there were no nuclear weapons. I mean, it goes back to long-term planning assumptions, and I don't really think it's racism. Let's take a concrete case. We have a lot of internal documents now, some interesting ones from the Nixon years. Nixon and Kissinger, when they were planning to overthrow the government of Chile in 1973, their position was that this government's intolerable, it's exercising its sovereignty, it's a threat to us, so it has to go.[14] It's what Kissinger called a virus that might spread contagion elsewhere, maybe into southern Europe—not that Chile would attack southern Europe—but that a successful, social democratic parliamentary system would send the wrong message to Spain and Italy. They might be inclined to try the same, it would mean its contagion would spread and the system falls apart. And they understood that, in fact stated that, if we can't control Latin America, how are we going to control the rest of the world? We at least have to control Latin America. There was some concern—which was mostly meaningless, but it was there—about a Soviet penetration into Latin America, and they recognized that if Europe gets more involved in Latin America, that would tend to deter any Soviet penetration, but they concluded the US couldn't allow that because it would interfere with US dominance of the region. So, it's not racist. It's a matter of dominance.

In fact, the same is happening with NATO. Why didn't NATO disappear after the Soviet Union collapsed? If anybody read the propaganda, they'd say, "Well, it should have disappeared, it was supposed to protect Europe from the Russian hordes." Okay, no more Russian hordes, so it should disappear. It expanded in violation of verbal promises to Gorbachev. And it expanded, I think, largely in order to keep Europe under control. One of the purposes of NATO all along was to prevent Europe from moving in an independent path, maybe a kind of Gaullist path, and they had to expand NATO to make sure that Europe stays a vassal. If you look back to the planning record during the Second World War, it's very instructive. It's almost never discussed, but there were high-level meetings from 1939 to 1945 under the Roosevelt administration, which sort of planned for the postwar years. They knew the United States would emerge from the war at least very well off and maybe completely triumphant. They didn't know how much at first. The principles that were established were very interesting and explicit, and later implemented. They devised the concept of what they called the Grand Area, which the US must dominate. And within the Grand Area, there can be no exercise of sovereignty that interferes with US plans—explicit, almost those words. What's the Grand Area? Well, at a minimum, it was to include the entire Western Hemisphere, the entire Far East, and the whole British Empire—former British Empire—which, of course, includes the Middle East energy resources. As one high-level advisor later put it: "If we can control Middle East energy, we can control the world."[15] Well, that's the Grand Area.

As the Russians began to grind down the German armies after Stalingrad, they recognized that Germany was weakened—at first, they thought that Germany would emerge from the war as a major power. So the Grand Area planning was extended to as much of Eurasia as possible, including at least Western Europe, which is the industrial-commercial center of the region. That's the Grand Area, and within that area, there can be no exercise of sovereignty. Of course, they can't carry it off.

For example, China is too big to push around and they're exercising their sovereignty. Iran is trying, it's small enough so you can push them around—they think so. Even Latin America is getting out of control. Brazil was not following orders. And, in fact, a lot of South America isn't, and the whole thing is causing a lot of desperation in Washington. You can see it if you look at the official pronouncements. China is not paying attention to US sanctions on Iran. US sanctions on Iran have absolutely no legitimacy. It's just that people are afraid of the United States. And Europe more or less goes along with them, but China doesn't. They disregard them. They observe the UN sanctions, which have formal legitimacy but are toothless, so they're happy to observe them. The major effect of the UN sanctions is to keep Western competitors out of Iran, so they can move in and do what they feel like. The US is pretty upset about it. In fact, the State Department issued some very interesting statements, interesting because of their desperate tone. They warned China that, this is almost a quote, "if you want to be accepted into the international community, you have to meet your international responsibilities, and the international responsibilities are to follow our orders." You can see both the desperation in US planning circles and you can kind of imagine the reaction of the Chinese foreign office, they're probably laughing, you know, why should they follow US orders? They'll do what they like.

They're trying to recover their position as a major world power. For a long time they were *the* major world power before what they call the "century of humiliation." They are now coming back to a three-thousand-year tradition of being the center of the world and dismissing the barbarians. So, okay, "we'll just go back to that and the US can't do anything about it," which is causing enormous frustration. That's why they get terribly upset when China doesn't observe US sanctions on Iran. By now it's not China and Iran that are isolated on Iran sanctions; it's the United States that's isolated. The nonaligned countries—118 countries, most of the world—have always supported Iran's right to enrich uranium, still do. Turkey recently constructed a pipeline to Iran, so has Pakistan. Turkey's trade with Iran has been going way up, they're planning to triple it the next few years. In the Arab world, public opinion is so outraged at the United States that a real majority now favors Iran developing nuclear weapons, not just nuclear energy. The US doesn't take that too seriously, they figure that dictatorships can control the populations. But when Turkey's involved or, certainly, when China's involved, it becomes a threat. That's why you get these desperate tones.

Apart from Europe, almost nobody's accepting US orders on this. Brazil's probably the most important country in the South. Not long ago, Brazil and Turkey made a deal with Iran for enriching a large part of the uranium; the US quickly shot that down. They don't want it, but the world is just hard to control.[16] The Grand Area planning was okay at the end of the Second World War when the US was overwhelmingly dominant, but it has been kind of fractured ever since—and during the last few years, considerably. And I think this is related to the proliferation issues. The US is strongly supporting India and Israel, and the reason is they've now turned India into a close strategic ally—Israel always was. India, on the other hand, is playing it pretty cool. They're also improving their relations with China.

President Obama recently secured military basing rights in Australia and formed a new free-trade pact, the Trans-Pacific Partnership, which excludes China. Is this move related to the South China Sea?

Yes, in particular that, but it's more general. It has to do with the "classic security dilemma" that I mentioned before, referring to the strategic analysis literature. China's efforts to gain some measure of control over nearby seas and its major trade routes are inconsistent with what the US calls "freedom of the seas"—a term that doesn't extend to Chinese military maneuvers in the Caribbean or even most of the world's oceans, but does include the US right to carry out military maneuvers and establish naval bases everywhere. For different reasons, China's neighbors are none too happy about its actions, particularly Vietnam and the Philippines, which have competing claims to these waters, but others as well.[17] The focus of US policy is slowly shifting from the Middle East—though that remains—to the Pacific, as openly announced. That includes new bases from Australia to South Korea (and a continuing and very significant conflict over Okinawa), and also economic agreements, called "free-trade agreements," though the phrase is more propaganda than reality, as in other such cases.[18] Much of it is a system to "contain China."

To what degree are current maritime sovereignty disputes related to oil and gas reserves?

In part. There are underseas fossil-fuel resources, and a good deal of contention among regional states about rights to them. But it's more than that. The new US base on Jeju Island in South Korea, bitterly protested by islanders, is not primarily concerned with energy resources. Other issues have to do with the Malacca Straits, China's main trade route, which does involve oil and gas but also much else.[19]

In the background is the more general concern over parts of the world escaping from US control and influence, the contemporary variant of Grand Area policies. Much of this extends the practice of earlier hegemonic powers, though the scale of US post–World War II planning and implementation has been in a class by itself because of its unique wealth and power.

5.

China and the Green Revolution

LARAY POLK: *In researching the cutting-edge innovations in energy in the US, it's pretty much the same players. GE; IBM; Raytheon; the DOE, they're funding fusion research; and a whole new department called ARPA-E based on DARPA, but with a focus on energy. Soldiers are currently using solar cells in the field and the navy is testing algae-based fuel.*[1]

NOAM CHOMSKY: Got to keep the military going, doesn't matter about the rest of us. Profits and the military: those are the two things that matter. And the military, of course, is not unrelated to profit.[2]

It's discouraging to see who is developing what. I can't see any of those entities marketing energy any differently than from the past.

It's even true of very simple things just take weatherization of homes. That's not high technology. It could put huge numbers of people to work, it would be a great stimulus for the economy, and it would be quite effective in retarding the effects of climate change. It's not a solution to the problem, but at least it provides more time to do something about it.

Recently there was a company in England that does weatherization that announced it had essentially done about everything you can in England. Almost everybody had it and they wanted to shift to the United States with a huge untapped market, but they're not sure it's going to be economically feasible here because they don't get any assistance.[3]

And, in fact, that's what's happening with the green technology. China provides a support system for development of green technology.[4] The United States does too, but a lot is for the support of military technology. That's actually a change from the past, a regression from the past. The actual US economy since the colonies has relied quite substantially on government intervention. That goes right back to the earliest days of independence, and for advanced industry in the latter part of the nineteenth century. The American system of mass production, interchangeable parts, quality control, and so on—which kind of astonished the world—was largely designed in government armories. The railroad system, which was the biggest capital investment and, of course, extremely significant for economic development and expansion, was managed by the Army Corps of Engineers. It was too complicated for private business.

Taylorism, the management technique that essentially turns workers into robots, came out of government and military production. The same was true of radio in the 1920s, but the big upsurge was in the postwar period—right where we are, in fact. Down below where we're sitting, there used to be a Second World War temporary building, where I was for many years, where they were developing—this was the 1950s, and then on through the '60s and beyond—computers, the beginnings of the Internet, information technology, software, everything that became the modern high-tech revolution. Almost all of it was on Pentagon funding—ARPA, it was then, now it's DARPA—or just the three armed services, and that laid the basis for the US high-tech industries, very substantially, and, in fact, often with a long delay. Even computers, the core of the modern economy, were being developed from around the early 1950s, mostly with government funding. They weren't commercially viable for about thirty years.

IBM, by the early '60s, was finally able to produce its own computer. They had learned enough at the government-funded laboratories to do it. It was the world's fastest computer at the time, but it was way too expensive for business, they couldn't buy it, so the government purchased it—I think it was for Los Alamos. In fact, procurement—and not just for computers, but also over the whole economy—has been a huge form of subsidy and it continues.[5]

That's the way the economy developed. Now we're screaming at China because they're doing much the same with green technology, and we're not doing it. We've regressed to the extent that this isn't done very much anymore, although it still is plenty. If you take a walk around MIT, you'll see big buildings of drug companies and genetic engineering. The reason is they're feeding off the government-funded ideas, technology, and development that's being done at the research labs and research universities like MIT. If you strolled around campus fifty years ago, you'd see small start-ups of electronic firms. Actually, on what's now Route 128, that became Raytheon, and a high-tech corridor.[6]

There's nothing new about this. This is the way economies develop. If there is an exception, I haven't come across it. British development was like this, based on huge state intervention from the early eighteenth century. The same is true for the US, Germany, France, Japan, the East Asian miracles—all of them—China, of course. Market systems don't yield fundamental innovation and development for obvious reasons. Innovation and development are long-term projects. They don't give you profits tomorrow. In fact, they give you costs. So the state takes it over; the taxpayer, in other words, pays for it. It's a system of essentially public subsidy, private profit. And it's called capitalism, but has little resemblance to capitalism.

And Koch Industries?

You can look right out the window, that's the Koch building.

So taxpayers' money goes in, and there's a mingling of industrial interests with university resources—resources meaning the intellectual capital—

And that's publicly funded, substantially—

Then innovation goes out but it's filtered through intellectual property rights—

Which is another form of government subsidy, and a major form. Take a look at the World Trade Organization rules. They've imposed patent conditions for the developing countries, which would have killed off industrial development in the rich countries if they had ever had to adhere to them.[7] The United States, for example, relied substantially on technology transfer—what's now called piracy—from England, which was more advanced. In fact, England did the same for more advanced technology from India and Ireland, and from more skilled workers from the Lowlands, Belgium, and Holland. We did it then to England, and other countries are trying to do it too, but they're barred from it by what are called free-trade rules, meaning: we protect what we want and we impose a market rigor on you.

There have been some good studies of this. Among the main beneficiaries of the World Trade Organization's rigorous patent restrictions are the pharmaceutical corporations. They claim that they need it for research and development. This was investigated carefully by, among others, Dean Baker, a very good economist. He went through the records and found that the corporations themselves fund only a minority of their own R&D, and that's misleading because it tends to be oriented toward the marketing side, copycat drugs, and so on. The basic funding comes from either the government or from foundations. He calculated that if funding for R&D for the big pharmaceutical corporations was raised to 100 percent public, and they were then compelled to sell their goods on the market, there'd be a colossal saving to consumers and no patent rights.[8] But that's unthinkable; anything that interferes with profit is unthinkable. It can't be discussed.

What role do politicians play in the distribution of federal funding for R&D, subsidy, and procurement?

Congress of course provides the funds, and the executive is deeply involved in decisions and implementation, with close involvement of industry lobbyists throughout. That aside, the decision makers in government have intimate ties with the corporate beneficiaries of subsidy and procurement in many other ways, ranging from campaign funding to privileged positions in the private sector if they play the game by the corporate rules.[9]

What factors in addition to massive investment have put China in the lead in green technology?

The business press and technology journals describe many factors, among them providing the required infrastructure. In the case of green technology, China began fairly modestly, and has been steadily advancing.

Take solar panels. China began manufacturing them in conventional ways and gained a large market share. A good deal of innovation and development comes directly out of manufacturing experience. This is not labor-intensive industry, so low labor costs were apparently not a major factor. Over time China has taken the lead in advanced solar panel technology, and now substantially dominates the international market. To illustrate how the US is falling behind in advanced manufacturing, US Secretary of Energy Steven Chu described Suntech power, after an on-site investigation, as a high-tech, automated factory that has developed a type of solar cell with world-record efficiencies. That is the result of careful planning in a framework of state industrial policy. It has its failures, but also real successes.[10]

6.

Research and Religion
(or, The Invisible Hand)

LARAY POLK: *Forty percent of the electorate in most states identify as evangelical. Pew Research indicates evangelical Christians largely reject anthropogenic climate change and are skeptical there is even solid evidence that the earth is warming.[1] So I think the extreme beliefs of the religious right benefit business interests and vice versa.*

NOAM CHOMSKY: That's an interesting combination because the business leadership tends to be secular. On social issues, they're what are called liberals. They're perfectly happy to mobilize and support, by what are world standards, extremist religious organizations as their sort of storm troops and they kind of have to do it. Take a look at recent American history: it's always been a very religious country, but until the past thirty years or so, there wasn't much political mobilization of the religious right. It took off pretty much in the '80s, and I think it's correlated with the fact that the Republicans, who were in the lead on this, began to take positions that are so hostile to interests of the public that they were going to lose any possible votes. They had to mobilize some kind of constituency, so they turned to what are called "social issues." The CEO of a corporation doesn't care that much if there's a law against, say, abortion. Their stratum of society is going to get it anyway, whatever the laws are. They'll have everything they want.

And if you have to throw some red meat to voters out there whose views you just think are ridiculous, then you do it. In a way, a most striking case is the environment. If you took a poll among CEOs of the major corporations that fund the Chamber of Commerce and so on, I suspect they would be just like faculty members at the university. Maybe they donate to the Sierra Club in their private lives, but not in their public roles. In their public roles, not only do they fund propaganda campaigns to undermine support for global warming, but they also support the political party which is mobilizing those efforts.[2] Quite an interesting split between an institutional role and what are probably private beliefs. In their institutional role they have a function: they must maximize short-term profit and market share. Their jobs and salaries depend upon it. And that institutional role is driving them toward what I suspect is a fairly conscious commitment to longer-term destruction.

Do you think those aligned with the Republican Party are mostly funding doubt—doubt that climate scientists can be trusted?[3]

Or anyone. In fact, if you look at polls now it's incredible. Last time I looked at a poll on this, I think approval of Congress was in single digits; the presidency, all corrupt, and Obama is probably anti-Christ anyway; the scientists, we can't trust them, pointy-headed liberals; banks, we don't like them, too big, but we're not going to do anything about it except fund them; and so on across the board. Trust in institutions is extremely low, and, unfortunately, that has some resonances rather similar to late Weimar Germany—plenty of differences, but there are some similarities that are worth concern.

And they can appeal to something quite objective. Take a look at post–Second World War history. The first two decades, the '50s and '60s, were periods of very substantial growth. In fact, the highest growth in the country's history—and egalitarian growth. People were gaining things, they were getting somewhere, and they had hope for the future and expectations, etc. The '70s was a transition period. Since the '80s, for the majority of the population, life has just gotten relatively worse: real wages and incomes have stagnated or declined; benefits, which were never very much, have declined; people have been getting by on working more hours per family, unsustainable debt, and asset inflation bubbles, but they crash.

So meanwhile, there's plenty of wealth around. If everything were impoverished, it wouldn't be so striking. You can read the front page of the *New York Times* and see it. A couple of weeks ago, they had an article on growing poverty in America, which is enormous, and another column on how luxury-good stores are marking up their prices because they can't sell them fast enough, might as well mark them up anyway. That's what the country's coming to look like, so people are angry—and rightly angry. And nothing is being done about it except to make it worse.

So it's a natural basis for preying on disillusionment and saying all institutions are rotten, get rid of all of them. The subtext being, you get rid of all of them, and we'll take control. Unfortunately that's the actual content of the libertarian conception, whatever the people may believe; they're effectively calling for corporate tyranny.

In The Protestant Ethic and the Spirit of Capitalism, *Max Weber wrote: "Absolute and conscious ruthlessness in acquisition has often stood in the closest connection with the strictest conformity to tradition."*[4] *Are there any parallels between Weber's observations in 1904 and present conditions?*

Depends what tradition one is thinking of. In the early days of the American Industrial Revolution, working people bitterly condemned the industrial system into which they were being driven as an assault on their fundamental values. They particularly condemned what they called "The New Spirit of the Age, Gain Wealth forgetting all but Self," that is, the doctrine of "absolute and conscious ruthlessness in acquisition."[5] The same was true of the people of England who resisted the enclosure movement and tried to preserve the "commons," which were to be the common property and source of sustenance for all, and to be cared for by all—also one of the core features of Magna Carta, long forgotten.[6] There are innumerable other examples illustrating the radical attack on tradition by the doctrines of ruthlessness in acquisition. I think Weber would have agreed.

Rick Santorum accused Obama of practicing "phony theology" related to radical environmentalists who have a worldview that elevates the "Earth above man." Santorum described his theology as "the belief that man should be in charge of the Earth and should have dominion over it and should be good stewards of it."[7] There seems to be a discrepancy in worldview as to what constitutes good stewardship.

Without speculating on what Santorum is talking about, let's take the lines you quote. A case can be made that the way to be a "good steward" of the earth is to abandon any thought of "dominion over it" and to recognize, with proper humility, that we must find a place within the natural world that will help sustain it not just for ourselves but for other creatures as well, and for future generations, recognizing values that are often upheld most firmly and convincingly within indigenous cultures.

Richard Land, host of the nationally syndicated radio show For Faith & Family, *said the Christian electorate "would love to see a false smarty pants decapitated by a real intellectual . . . He [Newt Gingrich] would tear Obama's head off."*[8] *He seems to be saying one type of intellectualism is acceptable and the right kind, but the other is not.*

When we look over the record of famous debates, we find that they are not "won" on the basis of serious argument, significant evidence, or intellectual values generally. Rather, the outcome turns on Nixon's five o'clock shadow, Reagan's sappy smile, lines like "have you no shame" or "you're no Jack Kennedy," etc. That's not surprising. Debates are among the most irrational constructions that humans have developed. Their rules are designed to undermine rational interchange. A debater is not allowed to say, "That was a good point, I'll have to rethink my views." Rather, they must adhere blindly to their positions even when they recognize that they are wrong. And what are called "skilled debaters" know that they should use trickery and deceit rather than rational argument to "win." I don't know who Richard Land is, and if he regards Gingrich as a "real intellectual," I don't see much reason to explore further.

The term "intellectual" is typically used to refer to those who have sufficient privilege to be able to gain some kind of audience when they speak on public issues. The world's greatest physicists are not called "intellectuals" if they devote themselves, laser-like, to the search for the Higgs boson. A carpenter with little formal schooling who happens to have very deep insight into international affairs and the factors that drive the economy and explains these matters to his family and friends is not called an "intellectual." There is evidence that the more educated tend to be more indoctrinated and conformist— but nevertheless, or maybe therefore, they tend to provide the recognized "intellectual class." We could devise a different concept that relates more closely to insight, understanding, creative intelligence, and similar qualities. But it would be a different concept.

Is there any value in skepticism without independent thinking?

Without independent thinking, skepticism would seem to reduce to "I don't accept what you say." It may be right not to accept it, but the stance is of value only if it is based on reasoned analysis and accompanied by sensible alternatives.

The current climate in the US—in addition to a lack of forums for reasoned debate—seems to be one of greed and also fear.

This has been a very frightened country from its origins. It's a striking feature of American culture that is interesting, well studied. Now, it's fear and also hopelessness. I'm just old enough to remember the Depression; objectively it was much worse. Most of my family was unemployed working-class, but there was a lot of hopefulness after the first few years. There was a sense that things are going to get better, we can do something about it, there's organizing and government efforts—it's bad, but we can get out of this. There isn't that feeling now, and it may be objectively right. If we continue on the path of financialization of the economy and offshoring of production, there's not going to be very much here for the working population.

It's kind of interesting if you look back at the classical economists, Adam Smith and David Ricardo. They were sort of aware of this—they didn't put it in precisely these terms—but if you take a look at Adam Smith's *The Wealth of Nations*, the famous phrase "invisible hand" appears once. It appears essentially in a critique of what's going on right now. What he pretty much says is that, in England, if merchants and manufacturers preferred to import from abroad and sell abroad, they might make profit, but it would be bad for England. He says they're going to have what sometimes is called a home bias—they'll prefer to do business at home, so as if by an invisible hand, England will be saved the ravages of a global market.[9]

David Ricardo was even stronger. He said that he knows perfectly well that his comparative advantage theories would collapse if English manufacturers, investors, and merchants did their business elsewhere, and he said he hopes very much that this will never happen—that they'll have, perhaps, a sentimental commitment to the home country—and he hopes this attitude never disappears. The insights of the classical economists were quite sound, whatever you think of the argument. And that's essentially the world we're living in.

7.

Extraordinary Lives

LARAY POLK: In your office, among all the reference materials, you have a rather large black-and-white photograph of Bertrand Russell. Did you have the opportunity to meet him?

NOAM CHOMSKY: We never met. Our only contact was in 1967, when we were about to issue the "Call to Resist Illegitimate Authority," advocating support for resistance, not just protest, to the Vietnam War. I was delegated to contact well-known figures to ask for their support. The first person I wrote to was Russell, who answered immediately, agreeing to sign the statement.

How much impact do you think Russell's nonproliferation work has had?[1]

It did not have as much of an impact as it should have. Russell was vilified in the US; there's a good account in the book *Bertrand Russell's America*.[2] Einstein, who often expressed similar views, was generally treated as a nice man who ought to go back to his study in Princeton. Nevertheless, it doubtless had some impact within those circles, then quite narrow, that were seeking to end the severe and immediate threat of nuclear weapons. In later years, that movement grew considerably, becoming a very powerful popular movement by the 1980s, probably a major factor in inducing Reagan to introduce his "Star Wars" fantasies so as to ward off protest. There's good work on this by Lawrence Wittner.[3]

Another scientist comes to mind, Linus Pauling, also a signatory to the Russell-Einstein Manifesto. I think you've mentioned having a great amount of respect for Pauling.

Pauling was a great scientist, but also a very dedicated and effective peace advocate. It was in the latter connection that I met him several times, on panels concerned with issues of war, aggression, and nuclear threats.

Also along these lines, you've mentioned Peggy Duff and her work with the Campaign for Nuclear Disarmament.[4]

Peggy Duff was a remarkable woman. In the late 1940s, she was active in trying to end Britain's shameful treatment of POWs after the war's end. She then became a leading figure in the CND, and soon went on to become the driving force in organizing the international movement of opposition to the Vietnam War, and also other crucial matters, such as the brutal denial of elementary rights to Palestinians. She organized international conferences, and much else, and also published very valuable and informative studies of ongoing events, bringing out a great deal of material that was missing or distorted in the general media.[5] By rights, she should have won the Nobel Peace Prize.

The statement you mentioned, "A Call to Resist Illegitimate Authority," was at the heart of a legal case in which you were named a co-conspirator. Is this the same incident of potential imprisonment that prompted your wife Carol to go back to school in case she had to become the sole breadwinner?[6]

Well before the trials were announced it was likely that the government would prosecute those who they regarded— mostly wrongly—as leaders of the resistance. That's why Carol went back to school after sixteen years (we had three kids to support). I was an unindicted co-conspirator in the first trial, but on the opening day the prosecuting attorney announced that I would be the primary defendant in the next trial—eliciting an objection from defense counsel. The reason why I was a co-conspirator and others were conspirators was comical, but in fact the entire government case was worthy of the Marx brothers, and provided some interesting insight into the incapacity of the political police to comprehend dissent and resistance.[7]

Pauling said of his nonproliferation work, "As scientists we have knowledge of the dangers involved and therefore a special responsibility to make these dangers known."[8] It seems that being honest with the science is not enough, that one has to also be engaged in international affairs and have a willingness to explore alternative definitions of what security means. Perhaps it could even be described as possessing a social direction that is different from the aspirations of politicians and others in the expert class.

It illustrates a basic moral principle. Privilege confers opportunity, and opportunity confers responsibility. Expert knowledge is one component of privilege. Politicians may sometimes have special knowledge, but that cannot be assumed.

Russell, Pauling, Duff, and others like them had integrity, and were willing to act in accord with decent values. In every society I know of since classical times there have been honest dissidents, usually a fringe, almost always punished in one or another way. The kind of punishment depends on the nature of the society. In contrast, obedience and subordination to power are typically honored within the society, even though often condemned by history (or in enemy states).[9]

In 1967 George Steiner wrote an open letter in reference to your essay, "The Responsibility of Intellectuals." His letter and your response were published together in the New York

Review of Books. *Is there anything memorable or significant to you about that exchange?*[10]

What is significant is that it took place. There was a good deal of soul-searching then, primarily among young people, about the course to follow as the Vietnam War moved on from major war crime to utter obscenity. And it reached to a certain extent to privileged intellectuals, the kind of people who read and wrote in the *Review*. One question—proper, and difficult—was whether to move on from protest to direct resistance, with all of its uncertainties and likely personal costs. Actually I'd been involved in it for several years before, in a tamer version: efforts to organize a national tax-resistance campaign in protest against the war. But by 1967, things were moving to a new stage.

What has changed and what has stayed the same since 1967?

One important change is that there have been a lot of victories, sometimes reaching to issues that were barely on the agenda not very long ago, like gay rights. And consciousness has greatly changed in many domains. Easy to list: rights of minorities, women, even rights of nature; opposition to aggression and terror; and much else.

It's instructive to look back to see the horrendous atrocities that were easily tolerated then, but not today. It's also instructive to look back at some of the dramatic moments of the '60s, for example Paul Potter's SDS speech at the first major mobilization in 1965, where he roused the crowd by declaring that the time had come to "name the system"; he couldn't go on to name it, though now there would be no such hesitation. He opened by saying that "most of us grew up thinking that the United States was a strong but humble nation, that involved itself in world affairs only reluctantly, that respected the integrity of other nations and other systems, and that engaged in wars only as a last resort."[11] Few young activists would say that now.

The achievements of the activism of the '60s and their aftermath leave a significant legacy: it is possible to go on to take up what was cut off then. The fate of the civil rights movement is worth remembering. In the standard version, it peaked in 1963 with the March on Washington and Martin Luther King's "I Have a Dream" speech. That's the usual focus of the rhetoric on MLK Day. But King didn't go home then. He went on to confront the burning issues of the day: the Vietnam War and the plight of the poor, with organizing in urban Chicago and elsewhere.[12] The luster quickly dimmed among Northern liberals. It's fine to condemn racist Alabama sheriffs, but state crimes and class issues are off-limits. Few remember King's speech in 1968, shortly before he was assassinated. He was in Memphis, Tennessee, supporting a strike of sanitation workers, and was intending to lead a March on Washington to found a movement of the poor and to call for meaningful legislation to address their plight.[13] The march took place, led by his widow, Coretta King. It passed through the sites of bitter struggle in the South and reached Washington, where the marchers set up a tent encampment, Resurrection City.[14] On orders of the most liberal administration since FDR, it was raided and destroyed by the police in the middle of the night, and the marchers were driven out of Washington.

The unfulfilled tasks remain, by now with new urgency after the disastrous economic policies of the past generation. And they can be undertaken from a higher plane.

Many of the old difficulties remain. Movements arise and grow and disappear leaving little organizational structure or memory. Most activism begins from almost zero. It also tends to be separated from other initiatives in a highly atomized society that is in some ways demoralized and frightened, despite its extraordinary wealth, privilege, and opportunities. And there are now questions of decent survival that cannot be shunted aside: the persistent danger of nuclear war, and the threat of environmental disaster, already approaching, and likely to become far more severe if we persist on our present course of denial.

8.

MAD

(Mutually Assured Dependence)

LARAY POLK: *Kumi Naidoo, the international executive director of Greenpeace, has been criticized for bringing a social agenda, not unlike King's, to the cause of environmental issues. Naidoo has said in response to his critics: "Ever since I came into this job, I've been accused of selling out, but I genuinely, passionately feel that the struggle to end global poverty and the struggle to avoid catastrophic climate change are two sides of the same coin. Traditional Western-led environmentalism has failed to make the right connections between environmental, social and economic justice. I came to the environmental movement because the poor are paying the first and most brutal impacts of climate change."*[1]

NOAM CHOMSKY: I presume that serious environmentalists would agree that saving whales does not get at the root of the problem, and that occupying oil rigs is at best a tactic undertaken to direct attention to deeper causes. On Naidoo, his approach seems to me fully justified, in other respects too. The poor who are (as usual) the victims suffering the most have also often been in the forefront of addressing the root problems. One striking example is the People's Summit in Bolivia, with its call for a Universal Declaration on the Rights of Mother Earth, an appeal voiced by indigenous people worldwide and a challenge to the predatory and lemminglike pursuit of short-term gain by the rich.[2]

Looking at Bolivia's ecology, it makes sense why they would have the most robust protections for nature: their glaciers are melting, and they're losing the ability to predict natural cycles of water distribution necessary for maintaining food crops. Those are conditions not unique to Bolivia or Andean glaciers, yet they're prepared to act.[3] What aspects of cultural practice prepare some communities for addressing ecological realities head-on? Conversely, what aspects of cultural practice impair—and perhaps immunize—other communities to ecological realities?

Looking at the ecology of the rich societies—the US for example—it also makes sense to move toward robust protections for nature. These past few months provide many warnings.[4]

There are many differences between Bolivia—the poorest country in South America—and the US, which by rights should be the richest country in world history, thanks to its unparalleled advantages.

One difference is that the major political force within Bolivia is the indigenous majority. Not only in Bolivia, but worldwide, indigenous communities ("first nations," "aboriginal," "tribal," whatever they call themselves) have been in the forefront of recognizing that if there is to be a hope of decent survival, we must learn to organize our societies and lives so that care for "the commons"—the common possessions of all of us—must become a very high priority, as it has been in traditional societies, quite often. The West too. It's rarely recognized that Magna Carta not only laid the basis for what became over centuries formal protection for civil and human rights, but also stressed the preservation of the commons from autocratic destruction and privatization—the Charter of the Forests, one of the two components of Magna Carta.[5]

In contrast, the US is a business-run society, to an extent beyond any other in the developed world. Enormous power lies in the hands of a highly class-conscious business elite, who, in Adam Smith's words, are the "principal architects" of policy and make sure that their own interests are "most peculiarly attended to" no matter how "grievous" the effects on others, including the people of their own society and their colonies (Smith's concern) and future generations (which must be our concern). In the contemporary United States there has been an increasing growth in the power of the ideology of short-term gains, whatever the consequences. The US business classes have been admirably forthright in announcing publicly their intention of running huge propaganda campaigns to convince the public to ignore the ongoing destruction of the environment, by now quite hard to miss even for the most blind. And these campaigns have had some effect on public opinion, as polls show.[6]

As for what "immunize [the culture] to ecological realities," in the US, it is useful to read the public pronouncements of the Chamber of Commerce (the main business lobby), the American Petroleum Institute, and other core components of the dominant business classes. Of course, that requires the contributions of the information and political systems, largely willing to line up in the same parade with only occasional hesitation.

During the Tar Sands Action in Washington, a spokesperson for the American Petroleum Institute told the press, "the protesters are really protesting jobs." What do you make of API's statement?[7]

The translation of the API statement to English is easy: "The Tar Sands Action is protesting an initiative that will severely harm local environments and accelerate the global rush to disaster—while putting plenty of bucks in our pockets for us to hoard or spend while we watch the ship sinking."

From what I know of the Tar Sands Action, it consists of people whose priorities are virtually the opposite of the API's. They want to maintain an environment in which people can live decent lives, to protect their grandchildren from disaster, and to create far more good jobs by using the ample resources available to develop a sustainable energy future while also rebuilding a decaying society and turning it to different and far more healthy directions.[8] But, admittedly, they have inadequate concern for the bulging profits of the super-rich and their desperate need to run the world to the ground.

The lack of serious media attention seems to me to fall into the normal pattern of downplaying the threat of global warming, along with general dislike of popular activism, which might revitalize democracy and threaten elite control. As for the former pattern, it is standard. Open the morning's paper and it is likely to be illustrated.

Today (August 17, 2012), for example, the press reports increasing reliance on Saudi oil, welcoming their increased production in response to US demands, but warning of a problem: dependence on foreign sources. Fortunately, the report continues, the problem is only temporary because we will soon have massive supplies from Canadian tar sands and expansion of drilling in the Gulf of Mexico— while also accelerating the race toward environmental catastrophe, a topic too insignificant to mention.[9]

On the "big twin threats of nuclear weapons and climate change" and the fallacy of a "limited nuclear war," activist and physicist Lawrence Krauss wrote: "Recent studies have concluded that even a limited nuclear exchange between

Pakistan and India, for example—involving perhaps 100 warheads—would significantly disrupt the global climate for at least a decade and would kick at least 5 million tons of smoke into the stratosphere. Estimates suggest this would potentially lead to the death of up to a billion people because of the effect of this smoke on global agriculture."[10]

Any concluding comments on the threat of nuclear war in a world already challenged by ecological collapse?

Sixty years ago President Eisenhower warned that "a major war would destroy the Northern Hemisphere."[11] Notwithstanding his warning, a few years later President Kennedy was willing to face his subjective probability of one-third to one-half of nuclear war to establish the principle that we have the right to ring the USSR with missiles and military bases, but they do not have the right to place their first missiles beyond its borders, in Cuba, then being subjected to a brutal terrorist attack that was scheduled to lead to invasion in the month when the missiles were secretly dispatched.[12] That was the essence of the issue. We escaped that time, but it was not the last. A decade later, in 1973, Henry Kissinger called a high-level nuclear alert to warn the Russians to keep their hands off when he was informing Israeli leaders that they could violate with impunity a cease-fire established under US-Russian auspices—so we have just learned from declassified documents.[13] Ten years later, Reaganite adventurism, probing Russian defenses at their borders, led a serious war scare as Russia feared an imminent nuclear attack.[14] There have been all too many cases of programmed missile attacks aborted by human intervention minutes before launch, and while we don't have Russian records, it's likely that their performance is worse. Right now President Obama is planning to establish an antimissile system—recognized on all sides to be a potential first-strike weapon—close to the Russian border, leading them to enhance their offensive weapons capacity.[15] According to the German press, Israel right now is loading nuclear-tipped missiles on the advanced new submarines that Germany has transferred to Israel, in the full knowledge that they are likely to be deployed in the Persian Gulf as part of the threat of escalated war against Iran.[16] And there is much more.

All of these crises can be mitigated or overcome. Many of the major barriers to doing so are right at home—a fortunate situation, because these are the factors that we can best hope to influence—hardly easy, but not impossible.

Those who choose to know, do know. The current issue of the journal of the American Academy of Arts and Sciences is devoted to the exciting prospects for science in the twenty-first century. The distinguished scientist who introduces the collection reviews these possibilities, adding, rather plaintively, "If we can manage to avoid total human disaster resulting from societal and environmental challenges (matters that in fact demand our most serious and immediate attention)."[17]

Bolivian *campesinos* understand.[18]

Appendix 1

Conversation between Gen. Groves and Lt. Col. Rea, August 25, 1945

On September 12, 1945, the New York Times published a front-page story by William L. Laurence, "U.S. Atom Bomb Site Belies Tokyo Tales." The story and the transcript below have a direct correlation: Laurence's report downplays radiation as the cause of death and suffering as a result of the atomic bombs dropped on Hiroshima and Nagasaki, and portrays symptoms described by the Japanese as propaganda meant to solicit sympathy. Laurence had been hired in March 1945 by the US War Department to write official statements and news stories; in 1946 he won a Pulitzer for a series of ten articles appearing in the New York Times on the "significance of the atomic bomb."

TOP SECRET

MEMORANDUM of Telephone Conversation between General Groves and Lt. Col. Rea, Oak Ridge Hospital, 9:00 a.m., 25 August 1945.

G: ". . . . which fatally burned 30,000 victims during the first two weeks following its explosion."

R: Ultra-violet—is that the word?

G: Yes.

R: That's kind of crazy.

G: Of course, it's crazy—a doctor like me can tell that. "The death toll at Hiroshima and at Nagasaki, the other Japanese city blasted atomically, is still rising, the broadcast said. Radio Tokyo described Hiroshima as a city of death. 90% of its houses, in which 250,000 had lived, were instantly crushed." I don't understand the 250,000 because it had a much bigger population a number of years ago before the war started, and it was a military city. "Now it is peopled by ghost parade, the living doomed to die of radioactivity burns."

R: Let me interrupt you here a minute. I would say this: I think it's good propaganda. The thing is these people got good and burned—good thermal burns.

G: That's the feeling I have. Let me go on here and give you the rest of the picture. "So painful are these injuries that sufferers plead: 'Please kill me,' the broadcast said. No one can ever completely recover."

R: This has been in our paper, too, last night.

G: Then it goes on: "Radioactivity caused by the fission of the uranium used in atomic bombs is taking a toll of mounting deaths and causing reconstruction workers in Hiroshima to suffer various sicknesses and ill health."

R: I would say this: You yourself, as far as radioactivity is concerned, it isn't anything immediate, it's a prolonged thing. I think what these people have, they just got a good thermal burn, that's what it is. A lot of these people, first of all, they don't notice it much. You may get burned and you may have a little redness, but in a couple of days you may have a big blister or a sloughing of the skin, and I think that is what these people have had.

G: That is brought out a little later on. Now it says here: "A special news correspondent of the Japs said that three days after the bomb fell, there were 30,000 dead, and two weeks later the death toll had mounted to 60,000 and is continuing to rise." One thing is they are finding the bodies.

R: They are getting the delayed action of the burn. For instance, at the Coconut Grove, they didn't all die at once, you know—they were dying for a month afterward.

G: Now then, he says—this is the thing I wanted to ask you

about particularly—"An examination of soldiers working on reconstruction projects one week after the bombing showed that their white corpuscles had diminished by half and a severe deficiency of red corpuscles."

R: I read that, too—I think there's something hookum [*sic*] about that.

G: Would they both go down?

R: They may, yes—they may, but that's awfully quick, pretty terrifically quick. Of course, it depends—but I wonder if you aren't getting a good dose of propaganda.

G: Of course, we are getting a good dose of propaganda, due to the idiotic performance of the scientists and another one who is also on the project, and the newspapers and the radio wanting news.

R: Of course, those Jap scientists over there aren't so dumb either and they are making a play on this, too. They evidently know what the possibility is. Personally, I discounted an awful lot of it, as it's too early, and in the second place, I think that a lot of these deaths they are getting are just delayed thermal burns.

G: You see what we are faced with. Matthias is having trouble holding his people out there.

R: Do you want me to get you some real straight dope on this, just how it affects them, and call you back in a bit?

G: That's true—that's what I want. Did you also see anything about the Geiger counter? It says that the fact that the uranium had permeated into the ground has been easily ascertained by using a Geiger counter and it has been disclosed that the uranium used in the atomic bomb is harmful to human bodies. Then it talks about this, which is just the thing that we thought—The majority of injured persons received burns from powerful ultra-violet rays and those within a two-kilometer radius from the center received burns two or three times, which, I suppose, is second or third degree. Those within three to four kilometers received burns to the extent that their skin is burned bright red, but if these burns are caused by ultra-violet, they hardly felt the heat at that time. Later, however, blisters formed resulting in dropsy.

R: That's why I say it's got to be a thermal burn.

G: Then they talk about the burned portions of the bodies are infected from the inside.

R: Well, of course, any burn is potentially an infected wound. We treat any burn as an infected wound. I think you had better get the anti-propagandists out.

G: We can't, you see, because the whole damage has been done by our own people. There is nothing we can do except sit tight. The reason I am calling you is because we can't get hold of Ferry and because I might be asked at any time and I would like to be able to answer. Did you see about the Army men who had received burns on reconstruction? "Examination of 33 servicemen, of whom 10 had received burns in reconstruction projects, one week after the bombing took place, showed those with burns had 3150 white corpuscles and others, who were apparently healthy, had 3800, compared to the ordinary healthy person who has 7,000 to 8,000." This is a drastic decrease. Comes over from Tokyo. On the other hand, servicemen with burns had only 3,000,000 red corpuscles and others apparently healthy had just a little bit more when compared to 4,500,000 to 5,000,000 in the ordinary healthy person." What is that measured by?

R: You go by cubic millimeters. I would say this right off the bat—Anybody with burns, the red count goes down after a while, and the white count may go down, too, just from an ordinary burn. I can't get too excited about that.

G: We are not bothered a bit, excepting for—what they are trying to do is create sympathy. The sad part of it all is that an American started them off.

R: Let me look it up and I'll give you some straight dope on it.

G: This is the kind of thing that hurts us—"The Japanese, who were reported today by Tokyo radio, to have died mysteriously a few days after the atomic bomb blast, probably were the victims of a phenomenon which is well known in the great radiation laboratories of America." That, of course, is what does us the damage.

R: I would say this: You will have to get some big-wig to put a counter-statement in the paper.

Source: National Security Archive

NOAM CHOMSKY
LINUS PAULING

DATE: TUESDAY, OCTOBER 10

TIME: 8:00 PM (COME EARLY TO BE ASSURED A SEAT)

PLACE: LOYOLA COLLEGE, MAIN AUDITORIUM (7141 Sherbrooke St. West,
near Montreal West Station)

ADMISSION: 50¢ DONATION

MASTER OF CEREMONY: PROFESSOR JEFFREY ADAMS

THE
VIETNAM
WAR–
WHAT IS TO BE DONE:

NOAM CHOMSKY - Presently, Professor of Modern Languages and
Linguistics, M.I.T.; Consulting Editor of Ramparts; Author
of books and articles on linguistics, also "The Responsibility
of Intellectuals" in New York Review of Books, Feb., 1967.

LINUS PAULING - Nobel Prize for Chemistry, 1954; Nobel Prize for
Peace, 1962; Taught in Chemistry Department, California Institute
of Technology (Pasedena), 1922-64; Research Professor, Center for
the Study of Democratic Institutions, 1963-67; Author of numerous
articles and books including No More War, 1958.

SPONSORED BY THE UNIVERSITIES COMMITTEE FOR PEACE IN VIETNAM. The UCPV was formed
in order to coordinate the efforts of that part of the academic community of
Montreal interested in working toward the goal of a peaceful solution to the
Vietnam War. Formed last March, its activities have included the holding of public
meetings, city-wide leafletting, aid to American war resisters, lobbying and the
support of anti-war demonstrations. Its members are drawn from the teaching and
professional staffs of Loyola College, McGill University, Sir George Williams
University and Universite de Montreal. Inquiries may be addressed to:
George Lermer, 5067 Bourassa, Pierrefonds, P.Q.

Appendix 2

Flyer for UCPV Event, October 10, 1967

This event, hosted by the academic community of Montreal, demonstrates the transnational involvement in resistance to the war in Vietnam. It was one of several events where Chomsky and Pauling were coparticipants.

NOAM CHOMSKY
LINUS PAULING

DATE: TUESDAY, OCTOBER 10

TIME: 8:00 PM (COME EARLY TO BE ASSURED A SEAT)

PLACE: LOYOLA COLLEGE, MAIN AUDITORIUM (7141 Sherbrooke St. West, near Montreal West Station)

ADMISSION: 50¢ DONATION

MASTER OF CEREMONY: PROFESSOR JEFFREY ADAMS

THE VIETNAM WAR—WHAT IS TO BE DONE?

NOAM CHOMSKY - Presently, Professor of Modern Languages and Linguistics, M.I.T.; Consulting Editor of *Ramparts*; Author of books and articles on linguistics, also "The Responsibility of Intellectuals" in New York Review of Books, Feb., 1967.

LINUS PAULING - Nobel Prize for Chemistry, 1954; Nobel Prize for Peace, 1962; Taught in Chemistry Department, California Institute of Technology (Pasadena), 1922–64; Research Professor, Center for the Study of Democratic Institutions, 1963–67; Author of numerous articles and books including *No More War*, 1958.

SPONSORED BY THE UNIVERSITIES COMMITTEE FOR PEACE IN VIETNAM. The UCPV was formed in order to coordinate the efforts of that part of the academic community of Montreal interested in working toward the goal of a peaceful solution to the Vietnam War. Formed last March, its activities have included the holding of public meetings, city-wide leafletting, aid to American war resisters, lobbying and the support of anti-war demonstrations. Its members are drawn from the teaching and professional staffs of Loyola College, McGill University, Sir George Williams University and Universite de Montreal. Inquiries may be addressed to: George Lermer, 5067 Bourassa, Pierrefonds, P.Q.

Source: Special Collections & Archives Research Center,
Oregon State University

Appendix 3

Scientists Condemn the Destruction of Crops in Vietnam, January 21, 1966

In response to a front-page article in the New York Times, "U.S. Spray Planes Destroy Rice in Vietcong Territory," a small group circulated a petition denouncing the destruction of food crops. The petition classifies the activity as indiscriminate chemical warfare and warns such practices would encourage other countries to employ similar tactics. Almost a year later, more than five thousand scientists signed a similar petition calling for a ban on chemical and biological weapons. One of the lead authors of both petitions, Matthew S. Meselson, studied with Linus Pauling at Caltech.

The use of crop-destroying chemicals by American forces in Vietnam was condemned this week in a statement by 29 scientists and physicians from Harvard, M.I.T., and several nearby institutions.

The statement made reference to a New York *Times*

dispatch which reported that, as part of a "large program of 'food denial' to the Vietcong," U.S. aircraft have been spraying rice crops with a "commercial weedkiller, identical with a popular brand that many Americans spray on their lawns." The *Times* report added that "it is not poisonous, and officials say that any food that survives its deadening touch will not be toxic or unpalatable."

The areas involved, according to the report, cover only a "small fraction—50,000 to 70,000 acres—of the more than eight million acres of cultivated land in South Vietnam." The program was reported "aimed only at relatively small areas of major military importance where the guerillas grow their own food or where the population is willingly committed to their cause." "Experience has shown," the *Times* stated, "that when the chemical is applied during the growing season, before rice and other food plants are ripe, it will destroy 60 to 90 percent of the crop."

John Edsall, Harvard professor of biochemistry, served as spokesman for the protesting group. The statement follows.

"We emphatically condemn the use of chemical agents for the destruction of crops, by United States forces in Vietnam as recently reported in the New York *Times* of Tuesday, 21 December 1965. Even if it can be shown that chemicals are not toxic to man, such tactics are barbarous because they are indiscriminate; they represent an attack on the entire population of the region where the crops are destroyed, combatants and non-combatants alike. In the crisis of World War II, in which the direct threat to our country was far greater than any arising in Vietnam today, our government firmly resisted any proposals to employ chemical or biological warfare against our enemies. The fact that we are now resorting to such methods shows a shocking deterioration of our moral standards. These attacks are also abhorrent to the general standards of civilized mankind, and their use will earn us hatred throughout Asia and elsewhere.

"Such attacks serve moreover as a precedent for the use of similar but even more dangerous chemical agents against our allies and ourselves. Chemical warfare is cheap; small countries can practice it effectively against us and will probably do so if we lead the way. In the long run the use of such weapons by the United States is thus a threat, not an asset, to our national security.

"We urge the President to proclaim publicly that the use of such chemical weapons by our armed forces is forbidden, and to oppose their use by the South Vietnamese or any of our allies."

The signers of the statement were as follows.

Harvard: John Edsall, Bernard Davis, Keith R. Porter, George Gaylord Simpson, Matthew S. Meselson, George Wald, Stephen Kuffler, Mahlon B. Hoagland, Eugene P. Kennedy, David H. Hubel, Warren Gold, Sanford Gifford, Peter Reich, Robert Goldwyn, Jack Clark, and Bernard Lown.

Massachusetts General Hospital: Victor W. Sidel, Stanley Cobb, and Herbert M. Kalckar.

M.I.T.: Alexander Rich, Patrick D. Wall, and Charles D. Coryell.

Brandeis: Nathan O. Kapland and William P. Jencks.

Amherst: Henry T. Yost.

Dartmouth: Peter H. von Hippel.

Tufts: Charles E. Magraw.

Also, Albert Szent-Györgyi, director of the Institute for Muscle Research, Woods Hole, and Hudson Hoagland, director of the Worcester Institute of Experimental Biology.

Source: Science

Appendix 4

Nelson Anjain's Open Letter to Robert Conard, April 9, 1975

From 1946 to 1958, the US Nuclear Weapons Testing Program conducted sixty seven nuclear detonations in the Marshall Islands. In 1956, Merril Eisenbud, director of the AEC Health and Safety Laboratory, outlined the benefits of studying a Marshallese population inhabiting an environment known to be radioactively contaminated: "[N]ow that Island is safe to live on but is by far the most contaminated place in the world and it will be very interesting to go back and get good environmental data, . . . While it is true that these people do not live, I would say, the way Westerners do, civilized people, it is nevertheless also true that these people are more like us than the mice."

April 9, 1975.
Rongelap Island,
Micronesia.
Dr. Robert Conard
Brookhaven National Laboratory
Upton, Long Island, New York 11790

Dear Dr. Conard,

I'm sorry I was not at home when you visited my island. Instead, I have spent the past few months travelling to Japan and Fiji learning about treatment of atomic bomb victims and about attempts to end the nuclear threat in the Pacific.

Since leaving Rongelap on the peace ship <u>Fri</u>, I have learned a great deal and am writing to you to clarify some of my feelings regarding your continued use of us as research subjects.

I realize now that your entire career is based on our illness. We are far more valuable to you than you are to us. You have never really cared about us as people—only as a group of guinea pigs for your government's bomb research effort. For me and for the other people on Rongelap, it is life which matters most. For you it is facts and figures. There is no question about your technical competence, but we often wonder about your humanity. We don't need you and your technological machinery. We want our life and our health. We want to be free.

In all the years you've come to our island, you've never treated us as people. You've never sat down among us and really helped us honestly with our problems. You have told people that the "worst is over", then Lekoj Anjain died. I don't know yet how many new cases you'll find during your current trip, but I am very worried that we will suffer again and again.

I'll never forget how you told a newspaper reporter that is was <u>our</u> fault that Lekoj died because we wouldn't let you examine us in early 1972. You seem to forget that it is your country and the people you work for who murdered Lekoj.

As a result of my trip, I've made some decisions that I want you to know about. The main decision is that we do not want to see you again. We want medical care from doctors who care about us, not about collecting information for the U.S. government's war makers.

We want a doctor to live on our island permanently. We don't need medical care only when it is convenient for you to visit. We want to be able to see a doctor when we want to. America has been trying to Americanize us by flying flags and using cast-off textbooks. It's about time America gave us the kind of medical care it provides its own citizens.

We've never really trusted you. So we're going to invite doctors from hospitals in Hiroshima to examine us in a caring way.

We no longer want to be under American control. As a

representative of the United States, you've convinced us that Americans are out to dominate others, not to help them. From now on, we will maintain our neutrality and independence from American power.

There will be some changes made. Next time you try to visit be prepared. Ever since 1972 when we first stood up to you, we've been aware of your motives. Now that we know that there are other people in the world who are willing to help us, we no longer want you to come to Rongelap.

Sincerely,
Nelson Anjain,
Magistrate.

NA: sc
cc: Hon. Gary Hart, U.S. Senate
       Hon. Phillip Burton, U.S. House
       Hon. Kurt Waldheim, Secretary-General, United
         Nations
       Hon. Ataji Balos, Congress of Micronesia

Source: Marshall Islands Document Collection, US Department of Energy

Appendix 5

Marshallese Medical Records in Hands of Gensuikin, July 27, 1976

Robert Conard's dispatch below seems to validate Nelson Anjain's concern that interest in the medical affairs of the Marshallese people revolved around tightly controlled record keeping. Most significantly, the letter discloses an alliance formed between those who were exposed to radiation through nuclear bombs and those exposed through nuclear weapons testing.

July 27, 1976
Dr. James L. Liverman
Assistant Administrator for Environment & Safety
Division of Biological and Environmental Research
Energy Research and Development Administration
Washington, D.C. 20545

Dear Jim,

On July 26th a Mr. Murakami, reporter for the Japanese

newspaper Asahi in Washington, D.C. called about a story he had received from Japan that some 66 of our Marshallese medical records had been copied and were in the hands of the leftist anti A-bomb group (Gensuikin) in Japan. It was his opinion that they would get a doctor or doctors to review them (presumably in criticism of our examinations) to be used at the anniversary meetings of the Hiroshima bomb next month. I asked how they had gotten the records and he suggested that it may have happened in conjunction with the recent visit to Japan of the two Rongelap young men though he also thought Japanese from that group may have visited the Marshall Islands. I told him we had nothing to hide, but were disturbed about the unethical nature of obtaining the records. I also said that the records in the Marshall Islands were not complete but that we had much more extensive records on all individuals on our examination list at Brookhaven. I answered several questions for him concerning our findings and treatment of the Rongelap people and outlined our examination programs (annual surveys, semi-annual hematology checks and quarterly visits by our resident physician stationed at Kwajalein). He asked why Japanese were barred from visiting the Islands. I told him the only incident I know of was the aborted visit of a Japanese "medical" team (mostly reporters) that had occurred in 1971 due to lack of proper credentials. I told him we had had Drs. H. Ezaki and I. Kumatori from Japan visit us on past surveys and suggested he contact them or the Radiation Effects Research Foundation if he wished to get bona fide Japanese medical opinions about our surveys.

I am sending a copy of this letter to Dr. LeRoy Allen at the Radiation Effects Research Foundation and request that he let us know about any Japanese publicity which may pertain to this matter.

Sincerely,
Bob [signed]
Robert A. Conard, M.D.

RAC: im
cc: Dr. LeRoy R. Allen
            Dr. Bond
            Dr. Cronkite

Source: Marshall Islands Document Collection, US Department of Energy

Appendix 6

Memorandum on Iraqi Use of Chemical Weapons, November 1, 1983

In October 1983 Iran began pressing for a UN investigation into Iraq's use of chemical weapons. US cables from this time period indicate personnel knew about Iraq's "almost daily use" of CW against Iranians and Kurds, and sought to deal with the problem behind the scenes prior to an official address by the UN. The cable below states immediate intervention is needed in order to maintain credibility regarding US policy "to halt CW use whenever it occurs."

United States Department of State
Washington, D.C. 20520
November 1, 1983

INFORMATION MEMORANDUM
                                    S/S

TO: The Secretary

FROM: PM—Jonathan T. Howe
SUBJECT: Iraqi Use of Chemical Weapons

We have recently received additional information confirming Iraqi use of chemical weapons. We also know that Iraq has acquired a CW production capability, primarily from Western firms, including possibly a U.S. foreign subsidiary. In keeping with our policy of seeking to halt CW use wherever it occurs, we are considering the most effective means to halt Iraqi CW use including, as a first step, a direct approach to Iraq. This would be consistent with the way we handled the initial CW use information from Southeast Asia and Afghanistan, i.e., private demarches to the Lao, Vietnamese and Soviets.

As you are aware, presently Iraq is at a disadvantage in its war of attrition with Iran. After a recent SIG meeting on the war, a discussion paper was sent to the White House for an NSC meeting (possibly Wednesday or Thursday this week), a section of which outlines a number of measures we might take to assist Iraq. At our suggestion, the issue of Iraqi CW use will be added to the agenda for this meeting.

If the NSC decides measures are to be undertaken to assist Iraq, our best chance of influencing cessation of CW use may be in the context of informing Iraq of these measures. It is important, however, that we approach Iraq very soon in order to maintain the credibility of U.S. policy on CW, as well as to reduce or halt what now appears to be Iraq's almost daily use of CW.

Drafted: PM/TMP: JLeonard
                11/01/83: ph. 632-4814
                WANG #2485P

Clearances: PM/TMP: PMartinez
                PM/P — RBeers
                PM/RSA — PTheros
                NEA — DTSchneider
                P — AKanter
                NEA/ARN: DLMack
                INR/PMA: DHowells

Source: National Security Archive

Appendix 7

Open Letter to Africa, December 12, 2011

On December 7, 2011, during the UN Climate Change Conference in Durban, South Africa, Sen. James Inhofe of Oklahoma delivered a video message aimed at the international delegation: "Today I'm happy to bring you the good news about the complete collapse of the global warming movement and the failure of the Kyoto process. . . . For the past decade, I have been the leader in the United States Senate standing up against global warming alarmism. . . . You should know that global warming skeptics everywhere wish we could be with you celebrating the final nail in the coffin on location in South Africa." Inhofe is the minority leader of the Environment and Public Works Committee. His top campaign contributors include Koch Industries (oil, chemicals, and forest-derived products); Murray Energy (coal); Devon Energy (oil and gas); Contran Corporation (chemicals, metals, and radioactive waste disposal); and Robison International (lobbyists for defense, nuclear energy, GE, and IBM).

US Senator's Statement At COP17 Disappointed US

We are writing as US citizens to express our grave disappointment about the views expressed by our government representatives at COP17. On December 7, US Senator James Inhofe delivered a video message to the Durban delegation which was ill-informed and mean-spirited.

We, like many others in the US, accept the consensual science on climate change: it is happening and people are suffering from water shortages, the acidification of the oceans, and extreme weather events.

The carbon load in the atmosphere, caused mainly by fossil fuel combustion, is too great and must be reduced. This reduction must begin before 2020.

While it is true that the US is a democracy, it is also true that Inhofe, who serves on a very powerful committee on environmental issues, continues to do the dirty work for industry.

Industry interests are the main impediment to any necessary movement on climate change which must happen on a global scale.

It is more accurate to say we have a democracy that uses free elections to put in place known obstructionists, and a media that disproportionally gives a forum to economically driven ideology over sound science.

JACK MIMS AND LARAY POLK
Dallas

Source: Mercury (South Africa)

Anjali Appadurai's Speech in Durban, December 9, 2011

On December 8, 2011, as US climate negotiator Todd Stern took the stage at the UN Climate Change Conference, Abigail Borah, a Middlebury College student, stood up from the audience and gave a short speech before being escorted out by security: "2020 is too late to wait. We need an urgent path to a fair, ambitious and legally binding treaty. You must take responsibility to act now, or you will threaten the lives of the youth and the world's most vulnerable. You must set aside partisan politics and let science dictate decisions." The day after Borah's speech, another student, Anjali Appadurai, addressed the delegation from the podium. Both speeches were met with applause.

AMY GOODMAN: A number of protests are being held today at the climate change conference to protest the failure of world leaders to agree to immediately agree to a deal of binding emissions cuts. Earlier today, Anjali Appadurai, a student at the College of the Atlantic in Bar Harbor, Maine, addressed the conference on behalf of youth delegates.

CHAIRPERSON: I'd now like to give the floor to Miss Anjali Appadurai with College of the Atlantic, who will speak on behalf of youth non-governmental organizations. Miss Appadurai, you have the floor.

ANJALI APPADURAI: I speak for more than half the world's population. We are the silent majority. You've given us a seat in this hall, but our interests are not on the table. What does it take to get a stake in this game? Lobbyists? Corporate influence? Money? You've been negotiating all my life. In that time, you've failed to meet pledges, you've missed targets, and you've broken promises. But you've heard this all before.

We're in Africa, home to communities on the front line of climate change. The world's poorest countries need funding for adaptation now. The Horn of Africa and those nearby in KwaMashu needed it yesterday. But as 2012 dawns, our Green Climate Fund remains empty. The International Energy Agency tells us we have five years until the window to avoid irreversible climate change closes. The science tells us that we have five years maximum. You're saying, "Give us 10."

The most stark betrayal of your generation's responsibility to ours is that you call this "ambition." Where is the courage in these rooms? Now is not the time for incremental action. In the long run, these will be seen as the defining moments of an era in which narrow self-interest prevailed over science, reason and common compassion.

There is real ambition in this room, but it's been dismissed as radical, deemed not politically possible. Stand with Africa. Long-term thinking is not radical. What's radical is to completely alter the planet's climate, to betray the future of my generation, and to condemn millions to death by climate change. What's radical is to write off the fact that change is within our reach. 2011 was the year in which the silent majority found their voice, the year when the bottom shook the top. 2011 was the year when the radical became reality.

Common, but differentiated, and historical responsibility are not up for debate. Respect the foundational principles of this convention. Respect the integral values of humanity. Respect the future of your descendants. Mandela said, "It always seems impossible, until it's done." So, distinguished delegates and governments around the world, governments of the developed world, deep cuts now. Get it done.

Mic check!

PEOPLE'S MIC: Mic check!

ANJALI APPADURAI: Mic check!

PEOPLE'S MIC: Mic check!

ANJALI APPADURAI: Equity now!

PEOPLE'S MIC: Equity now!

ANJALI APPADURAI: Equity now!

PEOPLE'S MIC: Equity now!

ANJALI APPADURAI: You've run out of excuses!

PEOPLE'S MIC: You've run out of excuses!

ANJALI APPADURAI: We're running out of time!

PEOPLE'S MIC: We're running out of time!

ANJALI APPADURAI: Get it done!

PEOPLE'S MIC: Get it done!

ANJALI APPADURAI: Get it done!

PEOPLE'S MIC: Get it done!

ANJALI APPADURAI: Get it done!

PEOPLE'S MIC: Get it done!

CHAIRPERSON: Thank you, Miss Appadurai, who was speaking on behalf of half of the world's population, I think she said at the beginning. And on a purely personal note, I wonder why we let not speak half of the world's population first in this conference, but only last.

AMY GOODMAN: That was a speech by Anjali Appadurai here in Durban at the U.N. climate change talks. Just after her speech, as you heard, she led a mic check from the stage, a move inspired by the Occupy Wall Street protests around the world. This is *Democracy Now!*, democracynow. org, *The War and Peace Report*. I'm Amy Goodman, as we broadcast live from Durban, South Africa. Back in a moment.

Source: Democracy Now!

Appendix 9

Point Hope Protest Letter to JFK, March 3, 1961

The Inupiat, one of the oldest continuous communities in North America, successfully protested Edward Teller's Project Chariot; a scheme to carve out an Alaskan harbor with nuclear explosives in the 1960s. The community faces similar risks today as Shell moves forward with its exploration for crude in the Chukchi and Beaufort seas. Critics say the extreme weather conditions of the Arctic and an inadequate oil-spill response plan is a disaster in the making.

Point Hope, Alaska
March 12, 1961

U.S. Atomic Energy Comission
San Francisco Operation Offices
2111 Bancroft Way
Berkley 4, California

16872

Dear sir;

He, the village health council and the residents of Point Hope would like to share with you our letter to the President of the United States, how we feel about the Chariot Project at Cape Thompson, Alaska.

Sincerely,
Alice Weber (sic.

Point Hope
Alaska
March 3, 1961

Mr. John F. Kennedy
President of the United States
Washington, D.C.

Dear Mr. President:

We the Health Council of Point Hope and the residents of the village don't like to see the blast at Cape Thompson. We want to go on record as protesting the Chariot Project because it is too close to our homes at Point Hope and to our hunting and fishing areas.

All the four seasons, each month, we get what we need for living. In December, January, February and even March, we get the polar bear, seals, tomcod, oogrook, walrus, fox and caribou. In March we also get crabs. In April, May and June, we hunt whales, ducks, seals, white beluga, and oogrook. In July we collect crow-bell eggs from Cape Thompson and Cape Lisburne and store them for the summer. In the summer we get some seals, oogrook, white beluga, fish, ducks, and caribou. In the middle of September many of our village go up Noohpuk River to stay for the fishing and caribou hunting until the middle of November. In November we get seals again and we used the seal blubber for our fuel. The hair seal skin we used for trading groceries from the store.

The ice we get for our drinking water during the winter is about twelve miles off from the village towards Cape Thompson. We melt snow also to drink and for washing. In spring, May and June we used ocean ice. In the summer we get our water from the village well.

We are concerned about the health of our children and the mothers-to-be after the explosion. We read about "the cumulative and retained isotope burden in man that must be considered." We also knew about strontium 90, how it might harm people if too much of it get in our body. We have seen the Summary Reports of 1960, National Academy of Sciences on " The Biological Effects of Atomic Radiation."

We are deeply concerned about the health of our people now and for the future that is coming. The signatures on page two accompanying this letter are the three of residents of the village of Point Hope who share this concern and wish to express their protest against Project Chariot.

I, David Frankson, President
of Point Hope Village Council
have approved and released this
letter of protest on the 6th
day of March 1961 at Point Hope
Alaska. *David Frankson*
   President of Point Hope ...

Sincerely yours,
Officers and members
of Point Hope Village
Health Council
...
...
...
...
...

...the un... signed are residents of Point
... ... share the concern of the rest ...
... ... to express our protest against ... ...
Projects

Elaine Frankson
Amos Lane
Eunice Lane
Leonie Lane
Barbara Lane
Beatrice Vincent
Annie Lingook
George Omnik
Frank Omnik
Enid Omnik
John Omnik
Her mark — X
Molly Koomalgoalcheak
Rosemary Timothy
Jimmy Killigivuk
Her mark — X
Kate Killigivuk
Leo Attungowruk
Jackey Attungowruk
Her mark — X
Minnie Attungowruk
Tily Oktollik
Donald Oktollik
Irma Oktollik
Calvin W. Oktollik
Helen H. Sage
Helen ...

Georgeann Ormittuk
Gretchen Tuckfield
Sarah Kinguk
Tillie Milligrock
Aaron Milligrock
Aggie Frankson
Andrew Frankson
Charlie B Tuckfield
Bertha Koonuk
Peter Thomovzak
Annie Koonrooyak
Daniel Lisbourne
Ella Lisbourne
John C. Oktoct
Molly Oktollik
Daniel Attungowruk
Kathleen Attungowruk
Bernard Nash
Kirk Quinp
Moses Melik
Chester Sevek
Mary Dirka
Laura Kinneeveauk
Hubert Kinneeveauk

Hilda Jones
Violet ...
Silas ...
Myra ...
William Lis...
Carl Omnik
Panee... Om...
Claudia ...
Her mark — X
Mabel Han...
Sr. Nicholas Hank
Sunshine Tuck...
Bert Tuck...
Sophie Tuckfield
Joseph ...
Reuben Tou...
Ruth Tou...
Eva Attungowruk
Delia ...
Raymond St...
Rose Ella ...
Carl Tuyrayk...
Daisy Oomittuk
Antonio Weyiouanna
Hilda ...
Rosemary ...
Billy ...

Point Hope
Alaska
March 3, 1961
Mr. John F. Kennedy
President of the United States
Washington, D.C.

Dear Mr. President:

We the Health Council of Point Hope and the residents of the village don't like to see the blast at Cape Thompson. We want to go on record as protesting the Chariot Project because it is too close to our homes at Point Hope and to our hunting and fishing areas.

All the four seasons, each month, we get what we need for living. In December, January, February and even March, we get the polar bear, seals, tomcod, oogrook [bearded seal], walrus, fox and caribou. In March we also get crabs. In April, May and June, we hunt whales, ducks, seals, white beluga, and oogrook. In July we collect crowbell eggs from Cape Thompson and Cape Lisburne and store them for the summer. In the summer we get some seals, oogrook, white beluga, fish, ducks, and caribou. In the middle of September many of our village go up Kookpuk River to stay for the fishing and caribou hunting until the middle of November. In November we get seals again and we need the seal blubber for our fuel. The hair seal skin we used for trading groceries from the store.

The ice we get for our drinking water during the winter is about twelve miles off from the village towards Cape Thompson. We melt snow also to drink and for washing. In spring, May and June we used ocean ice. In the summer we get our water from the village well.

We are concerned about the health of our children and the mothers-to-be after the explosion. We read about "the accumulative and retained isotope burden in man that must be considered." We also know about strontium 90, how it might harm people if too much of it got in our body. We have seen the Summary Reports of 1960, National Academy of Sciences on "The Biological Effects of Atomic Radiation."

We are deeply concerned about the health of our people now and for the future that is coming. The signatures on page two accompanying this letter are the names of residents of the village of Point Hope who share this concern and wish to express their protest against Project Chariot.

Sincerely yours,
Officers and members
of Point Hope Village
Health Council

Source: US Department of Energy

Appendix Acknowledgments

Appendix 1
Reprinted with the permission of the National Security Archive
National Security Archive Electronic Briefing Book No. 162, s.v.
  "Document 76"
http://www.gwu.edu/~nsarchiv/NSAEBB/NSAEBB162/76.pdf

Appendix 2
Reprinted with the permission of the Special Collections & Archives
  Research Center, Oregon State University
Flyer for a presentation by Noam Chomsky and Linus Pauling on
  Vietnam War, 1967

Appendix 3
Reprinted with the permission of AAAS
"Scientists Protest Viet Crop Destruction" from "Congress:
  Productive Year Is Seen Despite Vietnam," *Science* 151
  (January 1966): 309

Appendix 4
"Letter to R Conard, Subject: Treatment of Atomic Bomb Victims
  and Attempts to End the Nuclear Threat in the Pacific (Marshall
  Islands), April 9, 1975"
Marshall Islands Document Collection, Office of Health, Safety and
  Security, Department of Energy
http://www.hss.energy.gov/healthsafety/ihs/marshall/collection/
  data/ihp2/1976_.pdf

Appendix 5
"Letter to J L Liverman, Subject: RE Story of 66 of Marshallese
  Medical Records Had Been Copied and Were in the Hands of

the Leftist Anti-A Bomb Group (Gensuikin) in Japan, July 27, 1976"

Marshall Islands Document Collection, Office of Health, Safety and Security, Department of Energy

http://hss.energy.gov/healthsafety/ihs/marshall/collection/data/ihp1a/1383_.pdf

Appendix 6

Reprinted with the permission of the National Security Archive

National Security Archive Electronic Briefing Book No. 82, s.v. "Document 24"

http://www.gwu.edu/~nsarchiv/NSAEBB/NSAEBB82/iraq24.pdf

Appendix 7

Reprinted with the permission of the *Mercury* (South Africa)

"US Senator's Statement At COP17 Disappointed Us"

http://www.highbeam.com/doc/1G1-275270064.html

Appendix 8

Reprinted with the permission of *Democracy Now!*

"'Get It Done': Urging Climate Change Justice, Youth Delegate Anjali Appadurai Mic Checks U.N. Summit"

http://www.democracynow.org/2011/12/9/get_it_done_urging_climate_justice

Appendix 9

"Health Council of Point Hope to J. Kennedy, March 3, 1961, Document #16872"

Coordination and Information Center, US Department of Energy, Las Vegas, NV

Notes

Preface

1 "To the world's military leaders, the debate over climate change is long over. They are preparing for a new kind of Cold War in the Arctic, anticipating that rising temperatures there will open up a treasure trove of resources, long-dreamed-of sea lanes and a slew of potential conflicts." Eric Talmadge, "As Ice Cap Melts, Militaries Vie for Arctic Edge," Associated Press, April 16, 2012. Areas of future hostilities over oil include the Strait of Hormuz, South China Sea, and Caspian Sea basin. Michael T. Klare, "Danger Waters: The Three Top Hot Spots of Potential Conflict in the Geo-Energy Era," TomDispatch.com, January 10, 2012. On drilling in the Chukchi and Beaufort seas, see note 3, chap. 1.

2 In 2005, while deep-water drilling in Angola, an Exxon spokesperson said, "All the easy oil and gas in the world has pretty much been found. Now comes the harder work in finding and producing oil from more challenging environments and work areas." This is proved to be true as the new frontiers of unconventional oil (Arctic offshore, oil sands, oil shale, pre-salt deepwater, tight oil) involve extreme environmental risk in sensitive areas such as the boreal forest and the world's oceans. Based on BP's data, the estimated time span of the "world proved [oil] reserves" in meeting current demand is forty-six years. John Donnelly, "Price Rise and New Deep-Water Technology Opened Up Offshore Drilling," *Boston Globe*, December 11, 2005; Mark Finley, "The Oil Market to 2030—Implications for Investment and Policy," *Economics of Energy*

& *Environmental Policy* 1, no. 1 (2012): 28, doi:10.5547/2160-5890.1.1.4.

3   Christian Parenti, *Tropic of Chaos: Climate Change and the New Geography of Violence* (New York: Nation Books, 2011), 226.

Chapter 1: Environmental Catastrophe

1   Ley de Derechos de la Madre Tierra, Ley Nro. 071 (Estado Plurinacional de Bolivia December 21, 2010), http://www.gobernabilidad.org.bo/. See also, agenda for "Rights of Mother Earth: Restoring Indigenous Life Ways of Responsibility and Respect," International Indigenous Conference, Haskell Indian Nations University, Lawrence, Kansas, April 4–6, 2012.

2   Pres. Nixon advocated for an autonomous regulatory agency for antipollution programs upon entering office. In 1969 Congress passed the National Environmental Policy Act (NEPA); within a year, the Environmental Protection Agency had been established. At the signing of NEPA, Nixon remarked, "[T]he 1970s absolutely must be the years when America pays its debt to the past by reclaiming the purity of its air, its waters, and our living environment. It is literally now or never." "The Guardian: Origins of the EPA," EPA Historical Publication-1 (Spring 1992); Dennis C. Williams, "The Guardian: EPA's Formative Years, 1970–1973," EPA 202-K-93-002 (September 1993).

3   With approval from the Obama administration, Royal Dutch Shell began exploratory drilling in the Chukchi and Beaufort seas in summer 2012. However, the inability to respond to changing sea-ice conditions "underscores environmentalists' concerns that Arctic Ocean conditions are too unpredictable for safe drilling and that industry isn't up to the challenge." Companies with similar plans include ExxonMobil (in partnership with Russia's OAO Rosneft), ConocoPhillips, and Statoil ASA. Tom Fowler, "Shell Races the Ice in Alaska," *Wall Street Journal*, August 20, 2012. On activism in Alaska, see Appendix 9.

4   "In 2009, for the first time, the U.S. Chamber of Commerce surpassed both the Republican and Democratic National Committees on political spending. . . . Not long ago, the Chamber even filed a brief with the EPA urging the agency not to regulate carbon—should the world's scientists turn out to be right and the planet heats up, the Chamber advised, 'populations can acclimatize to warmer climates via a range of behavioral, physiological and technological adaptations.' As radical goes, demanding that we change our physiology seems right up there." Bill McKibben, "Global Warming's Terrifying New Math," *Rolling Stone*, August 2, 2012. Four major companies have pulled out of the Chamber over its stance on climate: Apple, Pacific Gas and Electric, PNM Resources, and Exelon. Nike resigned its board position. David A. Fahrenthold, "Apple Leaves U.S. Chamber over Its Climate Position," *Washington Post*, October 6, 2009.

5   Near the end of his presidential bid, Huntsman changed position. On August 18, 2011, Huntsman tweeted, "To be clear. I believe in evolution and trust scientists on global warming. Call me crazy." At the Heritage Foundation on December 6, 2011, he asserted, "there are questions about the validity of the science, evidenced by one university over in Scotland [*sic*] recently." Huntsman's remarks also coincided with an anonymous hacker's release of stolen e-mails from the University of East Anglia and COP17 proceedings in Durban, South Africa. Evan McMorris-Santoro, "Jon Huntsman's Climate Change Flip Flop Explained," TalkingPointsMemo. com, December 6, 2011; Justin Gillis and Leslie Kaufman, "New Trove of Stolen E-mails from Climate Scientists is Released," *New York Times*, November 22, 2011. On influence of Tea Party on Republican campaigns, see note 3, chap. 6.

6   At a rally in Florida, after Hurricane Irene narrowly bypassed the state, Michele Bachmann told the audience: "I don't know how much God has to do to get the attention of the politicians. We've had an earthquake; we've had a hurricane. He said, 'Are you going to start listening to me here?'" Along

similar lines, less than a month after the explosion on the Deepwater Horizon oil rig in the Gulf of Mexico, Gov. Rick Perry described the BP spill as an "act of God." Adam C. Smith, "Michele Bachmann Rally Draws over 1,000 in Sarasota, but Some Prefer Rick Perry," *Tampa Bay Times*, August 29, 2011; Peggy Fikac, "Perry Stands by 'Act of God' Remark about Spill," *Houston Chronicle*, May 5, 2010.

7   Hugo Chávez, "Chavez Address to the United Nations," CommonDreams.org, September 20, 2006. On US-Venezuela energy relations, see note 8, this chapter.

8   "By the 1950s, low-cost oil from abroad, even with a 10 percent tariff and added transportation costs, began to displace American oil in the home market. In 1958, the Eisenhower administration, under pressure from the Texas oil lobby, imposed quotas. These lasted fourteen years and further depleted U.S. Reserves. . . . In 1959, Venezuela offered to open its domestic market to U.S. exports in exchange for privileged access to the American oil market. When the United States rejected the offer and abrogated a 1939 reciprocal trade agreement, Venezuela approached Saudi Arabia, the largest and lowest cost producer, to join it in convening the founding conference of the Organization of Petroleum Exporting Countries (OPEC) in Baghdad in 1960. OPEC exploited favorable circumstances to raise oil prices fourfold in 1973 and 1974, tenfold by 1981." *Encyclopedia of Tariffs and Trade in U.S. History*, ed. Cynthia Clark Northrup and Elaine C. Prange Turney (Westport, CT: Greenwood, 2003), 1:286.

9   In 2008 Florida State University's economics department received a pledge of $1.5 million from the Charles G. Koch Charitable Foundation. In exchange, any new hires for a program promoting "political economy and free enterprise" must pass approval of a Koch-appointed advisory committee. Two other schools have similar arrangements: Clemson University and West Virginia University. The Koch foundation also provided millions to George Mason University for the establishment of the Mercatus Center—described by one

political strategist as "ground zero for deregulation policy in Washington." Kris Hundley, "Billionaire's Role in Hiring Decisions at Florida State University Raises Questions," *Tampa Bay Times*, May 10, 2011.

10 Tom Hamburger, Kathleen Hennessey, and Neela Banerjee, "Koch Brothers Now at Heart of GOP Power," *Los Angeles Times*, February 6, 2011.

11 "*Mother Jones* has tallied some 40 ExxonMobil-funded organizations that either have sought to undermine mainstream scientific findings on global climate change or have maintained affiliations with a small group of 'skeptic' scientists who continue to do so." Chris Mooney, "Some Like It Hot," *Mother Jones*, May/June 2005. ExxonMobil and the Koch brothers are also both large supporters of ALEC, a group of corporate lobbyists and lawmakers who meet at yearly lavish confabs and provide legislative boilerplate at the state level. See Beau Hodai, "Publicopoly Exposed: How ALEC, the Koch Brothers and Their Corporate Allies Plan to Privatize Government," *In These Times*, July 2011.

12 See Naomi Oreskes and Erik M. Conway, *Merchants of Doubt: How a Handful of Scientists Obscured the Truth on Issues from Tobacco Smoke to Global Warming* (New York: Bloomsbury Press, 2010); Peter J. Jacques, Riley E. Dunlap, and Mark Freeman, "The Organisation of Denial: Conservative Think Tanks and Environmental Scepticism," *Environmental Politics* 17 (June 2008): 349–85, doi:10.1080/09644010802055576.

Chapter 2: Protest and Universities

1 See Gary Milhollin's testimony, "U.S. Export Control Policy toward Iraq," C-SPAN Video Library (C-SpanVideo.org), October 27, 1992. On Iraqi scientists invited by the DOE to attend a symposium on detonation physics, see Martin Hill, "Made in the USA: How We Sold Secrets to Iraq That Helped Saddam Hussein Go Nuclear," *Mother Jones*, May/June 1991. See also Mark Clayton, "The Brains behind Iraq's Arsenal: US-Educated Iraqi Scientists May Be as Crucial to Iraq's Threat

as Its War Hardware," *Christian Science Monitor*, October 23, 2002.

2   As early as 1983, US officials knew about Iraq's "almost daily use" of chemical weapons against the Kurds and Iranians. National Security Archive Electronic Briefing Book No. 82, s.v. "Document 25." In 1990, almost a year after Reagan left office, the US Marine Corps released a manual reiterating Iran, not Iraq, had employed chemical weapons: "Blood agents were allegedly responsible for the most infamous use of chemicals in the war—the killing of Kurds at Halabjah. Since the Iraqis have no history of using these two agents—and the Iranians do—we conclude that the Iranians perpetrated this attack. It is also worth noting that lethal concentrations of cyanogen are difficult to obtain over an area target, thus the reports of 5,000 Kurds dead in Halabjah are suspect." Marine Corps Publications Electronic Library, s.v. "FMFRP 3-203," December 1990, 100. In 2002 Pres. Bush reversed position: "The Iraqi regime has plotted to develop anthrax and nerve gas and nuclear weapons for over a decade. This is a regime that has already used poison gas to murder thousands of its own citizens, leaving the bodies of mothers huddled over their dead children. . . . This is a regime that has something to hide from the civilized world." George W. Bush, State of the Union Address (January 29, 2002).

3   The Reagan-Zia alliance began when the US and its European and Arab allies sought to arm the mujahideen's "jihad against the Soviet Union" in Afghanistan. The CIA, in partnership with Pakistani and Saudi intelligence agencies, funded the arming and training of an estimated thirty-five thousand Islamic militants from forty-three Muslim countries in Pakistani madrassas between 1982 and 1990. In return, Reagan agreed not to question Zia's policies: torture, drug trafficking by the army, and Pakistan's nuclear weapons program. According to journalist Ahmed Rashid, "This global jihad launched by Zia and Reagan was to sow the seeds of al Qaeda and turn Pakistan into the world center of jihadism for the next two decades. .

. . Reagan was to severely compromise the U.S. stance on nuclear proliferation by declining to question Islamabad's development of nuclear weapons—as long as Zia did not embarrass Washington by testing them." *Descent into Chaos: The United States and the Failure of Nation Building in Pakistan, Afghanistan, and Central Asia* (New York: Viking, 2008), 9, 38–39.

4 According to CIA station chief in Pakistan in 1981, Howard Hart, his orders were to "Go kill Soviet soldiers." Hart responded, "Imagine! I loved it." Tim Weiner, *Legacy of Ashes: The History of the CIA* (New York: Doubleday, 2007), 384. When *Le Nouvel Observateur* asked Brzezinski about any regrets with the secret US involvement in Afghanistan, he replied: "Regret what? That secret operation was an excellent idea. It had the effect of drawing the Russians into the Afghan trap and you want me to regret it? The day that the Soviets officially crossed the border, I wrote to President Carter, essentially: 'We now have the opportunity of giving to the USSR its Vietnam war'. Indeed, for almost 10 years, Moscow had to carry on a war that was unsustainable for the regime, a conflict that brought about the demoralization and finally the breakup of the Soviet empire." David N. Gibbs, "Afghanistan: The Soviet Invasion in Retrospect," *International Politics* 37 (June 2000): 242.

5 Andrew Higgins, "How Israel Helped to Spawn Hamas," *Wall Street Journal*, January 24, 2009.

6 Mark Curtis, *Secret Affairs: Britain's Collusion with Radical Islam* (London: Serpent's Tail, 2010).

7 Reactor fuel that is less than 20 percent ^{235}U is classified as LEU or "non-weapon-useable low-enriched uranium." Fuel that is greater than 20 percent ^{235}U is classified as HEU or high-enriched uranium—"usually weapon-grade uranium (WgU) containing 90 percent or more ^{235}U." Frank von Hippel, "A Comprehensive Approach to Elimination of Highly-Enriched-Uranium from All Nuclear-Reactor Fuel Cycles," *Science & Global Security* 12 (November 2004): 138, doi:10.1080/08929880490518045. Iran's right to enrich

fuel remains central to current tensions: "Iran claims it needs the higher enriched uranium to produce fuel for the Tehran reactor that makes medical radioisotopes needed for cancer patients." Ali Akbar Dareini, "Iran Claims Two Steps to Nuclear Self-Sufficiency," Associated Press, February 15, 2012. The US provided Iran with the Tehran Research Reactor in 1967. The reactor, from its inception, is designed to operate on HEU. Sam Roe, "An Atomic Threat Made in America," *Chicago Tribune*, January 28, 2007.

8   Bryan Bender, "Potent Fuel at MIT Reactor Makes for Uneasy Politics," *Boston Globe*, December 29, 2009.

9   Ibid. In addition to training nuclear engineers, the MIT Reactor "is also a money-making enterprise, by radiating seeds used in prostate cancer treatments and by turning silicon into high-performance semiconductors for the hybrid car market."

10   Robert F. Barsky, *Noam Chomsky: A Life of Dissent* (Toronto: ECW Press, 1997), 140.

11   Nanoscale science and engineering is a rapidly emerging area of federally funded R&D with possible applications in materials, manufacturing, energy, defense, communications, and health care. Funding is administered through the National Nanotechnology Initiative (NNI), and supports fifteen agencies including the DOE, DOD, NSF, and NIH. The agencies comprise an infrastructure of more than ninety major interdisciplinary research and education centers. One of the centers, MIT's Institute for Soldier Nanotechnologies (ISN), works in partnership with the army and industrial collaborators Raytheon, DuPont, and Partners HealthCare. On ISN and Future Force Warrior, see note 2, chap. 5.

12   See Chomsky's deconstruction of Greenspan's "miracles of the market"—the Internet, computers, information processing, lasers, satellites, and transistors—in *Rogue States: The Rules of Force in World Affairs* (Cambridge, MA: South End Press, 2000), chap. 13. Nanotechnology is expected to yield the next frontier of market developments, utilizing familiar technology-transfer mechanisms: "Nano is huge, with pervasive benefits

for society, the economy, and national security . . . [it's] on par with electricity, transistors, the Internet, and antibiotics. How do you know nano is hot? The VC (venture capital) community has embraced it." Lauren J. Clark, "ISN Director Ned Thomas Speaks on the Promises and Challenges of Nanotechnology," *ISN News*, February 2005, 6–7.

13   Chomsky's early technical reports bear the imprint of MIT's Research Laboratory of Electronics. RLE was founded in 1946 as a successor to the Radiation Laboratory (RadLab) developed during wartime. The RadLab produced nearly half of the radar used in World War II; one prototype is on view upon entering the building where Chomsky's office is located.

14   See Michael Albert, *Remembering Tomorrow: From SDS to Life After Capitalism, A Memoir* (New York: Seven Stories Press, 2007).

15   Vera Kistiakowsky, former MIT professor of physics, has expressed similar views: "Universities should not solicit or encourage funding by mission-oriented sources [e.g., Department of Defense] without a faculty consensus that this is desirable. Individual faculty members should take responsibility for foreseen consequences of their research, including those attached to seeking or accepting support from particular sources. Social responsibility should become important among the criteria of excellence at the universities, a factor in promotion and tenure decisions." "Military Funding of University Research," *Annals of the American Academy of Political and Social Science* 502 (March 1989): 153, doi:10.117 7/0002716289502001011.

Chapter 3: Toxicity of War

1   In 1969 Henry Kissinger said of the inhabitants of the Marshall Islands, "There are only 90,000 people out there. Who gives a damn?" Quoted in Jane Dibblin, *Day of Two Suns: U.S. Nuclear Testing and the Pacific Islanders* (New York: New Amsterdam Books, 1990). On contemporary life of the Marshallese, see André Vltchek,

"From the Kwajalein Missile Range to Fiji: The Military, Money and Misery in Paradise," *Asia-Pacific Journal* (October 2007).

2   "The principle victims of British policies are Unpeople—those whose lives are deemed worthless, expendable in the pursuit of power and commercial gain. They are the modern equivalent of the 'savages' of colonial days, who could be mown down by British guns in virtual secrecy, or else in circumstances where the perpetrators were hailed as the upholders of civilisation." Mark Curtis, *Unpeople: Britain's Secret Human Rights Abuses* (London: Vintage, 2004), 2. See also George Orwell's use of the term "unperson" in *1984*.

3   When Dr. Helen Caldicott was asked whether she thought this description was apt, she responded, "I would describe it as nuclear war without the blast, the effects of which will be endless." E-mail correspondence, February 16, 2012.

4   A joint project between the National Institute of Environmental Health Sciences (NIEHS) and Vietnam to study possible links between Agent Orange and health and environmental degradation never got off the ground. The study "was expected to provide evidence for a class action suit on behalf of millions of Vietnamese plaintiffs against US manufacturers of Agent Orange." Declan Butler, "US Abandons Health Study on Agent Orange," *Nature* 434 (April 2005): 687, doi:10.1038/434687a. On outcome of suit, see note 12, this chapter.

5   Fred A. Wilcox, *Scorched Earth: Legacies of Chemical Warfare in Vietnam* (New York: Seven Stories Press, 2011); *Waiting for an Army to Die: The Tragedy of Agent Orange*, 2nd ed. (New York: Seven Stories Press, 2011).

6   Samira Alaani, Muhammed Tafash, Christopher Busby, Malak Hamdan, and Eleonore Blaurock-Busch, "Uranium and Other Contaminants in Hair from the Parents of Children with Congenital Anomalies in Fallujah, Iraq," *Conflict and Health* 5 (September 2011): 1–15, doi:10.1186/1752-1505-5-15.

7   Patrick Cockburn, "Toxic Legacy of US Assault on Fallujah 'Worse than Hiroshima,'" *Independent* (London), July 24, 2010; Chris Busby, Malak Hamdan, and Entesar Ariabi,

"Cancer, Infant Mortality and Birth Sex-Ratio in Fallujah, Iraq 2005–2009," *International Journal of Environmental Research and Public Health* 7 (July 2010): 2828–37, doi:10.3390/ijerph7072828.

8   See Mads Gilbert and Erik Fosse, *Eyes in Gaza* (London: Quartet Books, 2010).

9   The DU penetrator was developed by metallurgist and engineer Paul Loewenstein (ca. 1958). He worked as technical director and vice president of Nuclear Metals, Inc. (NMI) from 1946 to 1999. Prior to becoming a privately owned business, NMI operated on the MIT campus in the Hood Building. In 1943 MIT had been designated a Manhattan Engineering District, producing alloys from ^{235}U and beryllium. In 1958 the operation, including machinery, staff, and licenses for uranium and beryllium, changed to private hands and relocated to Concord, MA. Renee Garrelick, *M.I.T. Beginnings: The Legacy of Nuclear Metals, Inc.* (Concord, MA: Nuclear Metals, 1995). MIT demolished the Hood Building due to contamination, and in the late 1990s, at the urging of citizens' groups, the NMI site in Concord was investigated for groundwater contamination. It was eventually placed on the EPA's National Priorities List; remediation continues into the present with an estimated cost of $63.9 million.

10 Wilcox, *Scorched Earth*, 124–31.

11 Official records claim Pres. Kennedy approved a program "to participate in a selective and carefully controlled joint program of defoliant operations in Viet Nam . . . proceeding thereafter to food denial only if the most careful basis of resettlement and alternative food supply has been created," on November 30, 1961. William A. Buckingham Jr., *Operation Ranch Hand: The Air Force and Herbicides in Southeast Asia 1961–1971* (Washington, DC: Office of Air Force History, 1982), 21. Other records indicate the decision to destroy crops had been made earlier in the month. On November 11, the NSC authorized the transport of "Aircraft, personnel and chemical defoliants" to Vietnam to "kill Viet Cong food crops." By November 27,

"spraying equipment had been installed on Vietnamese H-34 helicopters" and was "ready for use against food crops." George McT. Kahin, *Intervention: How America Became Involved in Vietnam,* 1st ed. (New York: Knopf, 1986), 478. On opposition to crop destruction, see Appendix 3.

12 In 1984 Monsanto and six other manufacturers settled with US veterans in a class-action lawsuit; $180 million was distributed according to a plan partially designed by US District Judge Jack B. Weinstein. In 2005 Weinstein denied claims sought by Vietnamese victims of Agent Orange on grounds of specific intent: "The United States did not use herbicides in Vietnam with the specific intent to destroy any group. Nor were those herbicides designed to harm individuals or to starve a whole population into submission or death. The herbicides were primarily applied to plants in order to protect troops against ambush, not to destroy a people." Vietnam Association for Victims of Agent Orange/Dioxin v. Dow Chemical Co. et al., MDL No. 381, 04-CV-400 (E.D.N.Y. March 25, 2005). See also Dominic Rushe, "Monsanto Settles 'Agent Orange' Case with US Victims," *Guardian* (London), February 24, 2012.

13 Martin Chulov, "Iraq Littered with High Levels of Nuclear and Dioxin Contamination, Study Finds," *Guardian* (London), January 22, 2010; Aseel Kami, "Iraq Scarred by War Waste," *Globe and Mail* (Toronto), October 24, 2008. Burn pits are another source of lethal toxicity: "Since 2003, defense contractors have used burn pits at a majority of U.S. military bases in Iraq and Afghanistan as a method of destroying military waste. The pits incinerate discarded human body parts, plastics, hazardous medical material, lithium batteries, tires, hydraulic fluids, and vehicles. Jet fuel keeps pits burning twenty-four hours a day, seven days a week." J. Malcolm Garcia, "Toxic Trash: The Burn Pits of Iraq and Afghanistan," *Oxford American*, August 24, 2011.

14 In 1994 Pres. Clinton established the Advisory Committee on Human Radiation Experiments (ACHRE) to investigate US government-funded research conducted between 1944

and 1974. A host of documents were sought, assembled, and declassified, establishing nearly four thousand radiation experiments involving plutonium and other atomic bomb materials; nontherapeutic research on children; total body irradiation; research on prisoners; intentional radioisotope distribution and atmospheric releases; and observational research involving uranium miners and residents of the Marshall Islands. In 1995 documents from the original ACHRE site were obtained by the National Security Archive (an independent nongovernmental research institute and library) located at George Washington University in Washington, DC.

Chapter 4: Nuclear Threats

1   In January 1995 Russia misidentified a Norwegian weather rocket as a US submarine-launched ballistic missile. Pres. Boris Yeltsin had the controls for a nuclear launch in hand, but decided at the last minute it had been a false alert. "As Russian capabilities continue to deteriorate, the chances of accidents only increase. . . . Russia's early warning systems are 'in a serious state of erosion and disrepair,' making it all the more likely that a Russian president could panic and reach a different conclusion than Yeltsin did in 1995." Joseph Cirincione, *Bomb Scare: The History and Future of Nuclear Weapons* (New York: Columbia University Press, 2007), 96–97.

2   UN Security Council Resolution 1887 was unanimously approved on September 24, 2009. Two days later, in Pittsburgh, PM Manmohan Singh told the press Pres. Obama had assured him the resolution—which calls on nonmembers of the NPT to join—wasn't aimed at India, and that the "US commitment to carry out its obligations under the civil nuclear agreements . . . remains undiluted." On October 2 Israeli officials said Obama had reassured them that the four-decade-old ambiguity policy that allows "Israel to keep a nuclear arsenal without opening it to international inspections" remained in effect. "NPT Resolution Not Directed against India: US," Indo-Asian News

Service, September 26, 2009; Eli Lake, "Obama Agrees to Keep Israel's Nukes Secret," *Washington Times*, October 2, 2009.

3   On Reagan-Zia alliance and nuclear program, see note 3, chap. 2.

4   James Lamont and James Blitz, "India Raises Nuclear Stake," *Financial Times*, September 27, 2009.

5   Prior to Pres. Bush's nuclear deal with India in 2006, the Nuclear Suppliers Group (NSG) functioned as a "relatively effective nonproliferation cartel." The NSG grew out of a secret meeting between US officials and nuclear technology suppliers in 1975 in response to India's first nuclear detonation a year earlier. The meeting established controls on the sale of flagged items and an agreement to bar sales to non-nuclear-weapon states for use at sites outside the purview of IAEA inspection. Bush allowed the selling of nuclear reactors, fuel, and technology to a country that maintains at least eight facilities kept off-limits to inspection. As a consequence, the move compromised a secondary system—the NPT as primary—of nonproliferation checks and balances. Cirincione, *Bomb Scare*, 37–38.

6   During the IAEA General Conference in 2009, back-to-back resolutions were adopted pertaining to the NPT and the establishment of a nuclear-weapon-free zone in the Middle East. Resolution 16 addresses the Middle East in general; it passed by a vote of 103–0. Resolution 17 specifically addresses Israel. It passed by a narrow margin of 49–45 with the "vote split among Western and developing nation lines." Upon passage, chief Israeli delegate David Danieli told the chamber, "Israel will not cooperate in any matter with this resolution." IAEA General Conference, GC(53)/RES/16 and RES/17, September 2009; Sylvia Westall, "UN Body Urges Israel to Allow Nuclear Inspection," Reuters, September 18, 2009.

7   A nuclear-weapon-free zone (NWFZ) in the Middle East was first proposed in 1962 by a group of Israeli intellectuals, the Committee for the Denuclearization of the Middle East, followed by a joint Egyptian-Iranian General Assembly resolution in 1974. The resolution has passed every year since, though numerous obstacles have prevented the zone from

being enacted. "Scientists Call for Nuclear Demilitarization in the Region," *Ha'aretz* (Hebrew), July 25, 1962; Nabil Fahmy and Patricia Lewis, "Possible Elements of an NWFZ Treaty in the Middle East," *Disarmament Forum*, no. 2 (2011): 39–50. On the US and collapse of the 2012 Helsinki Conference, see Noam Chomsky, "The Gravest Threat to World Peace," *Truth-Out.org*, January 4, 2013.

8   In December 1960 the US government submitted five questions to Israel regarding its possible nuclear weapons program: "(1) What are present GOI (government of Israel) plans for disposing of plutonium which will be bred in new reactor? (2) Will GOI agree to adequate safeguards with respect to plutonium produced? (3) Will GOI permit qualified scientists from the IAEA or other friendly quarters visit new reactor? If so, what would be earliest time? (4) Is a third reactor in either construction or planning stage? (5) Can Israel state categorically that it has no plans for developing nuclear weapons?" Avner Cohen, *Israel and the Bomb* (New York: Columbia University Press, 1998), 93–94.

9   On September 26, 1969, a policy of Israeli nuclear ambiguity was agreed upon by Pres. Nixon and PM Golda Meir. It remained secret until revealed by journalist Aluf Benn in 1991. Avner Cohen and Marvin Miller, "Bringing Israel's Bomb Out of the Basement: Has Nuclear Ambiguity Outlived Its Shelf Life?," *Foreign Affairs*, September/October 2010. Cohen has drawn parallels between Israel's policy and Iran's possible nuclear ambitions: "It is straddling the line, and in my opinion, Iran wants to, and can, remain for some time with the status of a state that might or might not have the bomb. Iran is a state of ambiguity." Noam Sheizaf, "Clear and Present Danger," *Ha'aretz*, October 29, 2010.

10   Louis Charbonneau, "U.S. and Other Big Powers Back Mideast Nuclear Arms Ban," Reuters, May 5, 2010.

11   Diego Garcia is home to one of five monitoring stations used to operate NAVSTAR GPS. Other terrestrial locations in the network include Hawaii, Colorado, Ascension Island,

and Kwajalein Atoll. Implemented by the Department of Defense in 1973, NAVSTAR GPS is a radio-navigation system that utilizes satellite to ground triangulation to provide precise geospatial coordinates for both military and civilian use (i.e., vehicle and cellular map locator). The system is usually referred to in its shortened form, GPS. "The Global Positioning System," National Academy of Sciences, 1997. The permanent eviction of the inhabitants of Diego Garcia (ca. 1973) to make way for US military operations is a continued issue of contention. Recent news of the UK government's plan to create a marine protection area has further inflamed the issue; a leaked diplomatic cable confirmed suspicions it was a move to deny Chagossians the right of return: "BIOT's former inhabitants would find it difficult, if not impossible, to pursue their claim for resettlement on the islands if the entire Chagos Archipelago were a marine reserve." *WikiLeaks*, s.v. "Cable 09LONDON1156, HMG Floats Proposal for Marine Reserve Covering," May 2009. On GPS and Kwajalein Atoll, see Vltchek, note 1, chap. 3.

12 In 2009 the Pentagon sent an "urgent operational need" funding request to Congress to fast-track the development and testing of the Massive Ordnance Penetrator (MOP), a thirty-thousand-pound bunker-busting bomb designed to hit underground targets. It was listed inside a ninety-three-page "reprogramming" request with hundreds of other items, and approved with little fanfare. Jonathan Karl, "Is the U.S. Preparing to Bomb Iran?," ABC News, October 6, 2009. On shipment of bunker-busting bombs to Diego Garcia, see Rob Edwards, "Final Destination Iran?," *Herald* (Scotland), March 14, 2010.

13 John J. Kruzel, "Report to Congress Outlines Iranian Threats," AFPS (Defense.gov), April 20, 2010.

14 See National Security Archive Electronic Briefing Book No. 255, "New Kissinger 'Telcons' Reveal Chile Plotting at Highest Levels of U.S. Government."

15 Under the leadership of Mohammed Mossadeq, Iran sought "increased benefits" from its resources, including a nationalized oil industry. A. A. Berle, former advisor to FDR, sent a dispatch to a friend in the State Department, stressing access to Persian Gulf oil translated to "substantial control of the world," and suggested an "appropriate formula" would be needed for intervention. The CIA launched Operation Ajax, staging acts of provocation aimed at ousting Mossadeq. In a coup on August 19, 1953, the goal was achieved: "The full consequences of that 1953 day when the shah was ushered back from exile have, indeed, never ended either for Iranians or Americans. For a quarter century [Shah] Reza Pahlavi controlled both his country and, it can be argued, American policy. He became one of the biggest customers ever for American military products, . . . Iran now became not simply an oil-bearing state, but also an aid in redressing other Cold War dilemmas, especially the efforts to overcome Vietnam-induced economic problems." Lloyd C. Gardner, *Three Kings: The Rise of an American Empire in the Middle East After World War II* (New York: New Press, 2009), 96–132.

16 "Iran Says Uranium to Go to Turkey, Brazil for Enrichment," Voice of America, May 17, 2010; David E. Sanger and Michael Slackman, "U.S. Is Skeptical on Iranian Deal for Nuclear Fuel," *New York Times*, May 17, 2010.

17 Journalist Zoher Abdoolcarim describes the disputes as a complex of regional relationships: "When it comes to feuds in the Pacific over islands and what lies beneath, it's not simply a case of China against everyone else. Depending on the dispute, it's also South Korea vs. Japan, Japan vs. Taiwan, Taiwan vs. Vietnam, Vietnam vs. Cambodia and numerous other permutations—for many of the same reasons supposedly behind China's actions. Resource grab. Patriotic posturing. Historical baggage (mostly to do with Japan's brutal occupation of most of East Asia before and through World War II). . . . Amid East Asia's island fever, there's big and small, strong and weak, rich and poor, and enlightened

and unenlightened self-interest. But not as innocent as good vs. evil." "Why Asia's Maritime Disputes Are Not Just about China," TIMEWorld (World.Time.com), August 19, 2012.

18   On tensions between free-trade agreements and the "higher values of the protection of the earth and people's livelihoods," see Vandana Shiva, *Stolen Harvest: The Hijacking of the Global Food Supply* (Cambridge, MA: South End Press, 2000). On the export of carbon dependency, see Bharti Chaturvedi, "Debate over FDI in Retail, Durban Talks Are Linked," *Hindustan Times*, December 4, 2011.

19   Jeju Island, located in the Korean Strait, is being prepared as "an expansive base which would be home to 20 warships and submarines and would serve as a strategic component in the U.S. military's sea-based ballistic missile defense system." In July 2012, the Indian Navy announced a new base on the Andaman and Nicobar Islands as a means to patrol the Malacca Straits, one of the busiest sea-lanes in the world. In 2006, the *Hindustan Times* reported plans for the base came from the US, "given its growing comfort level with India and their growing strategic relationship, [the US] is delegating to New Delhi a role that would have been unthinkable even a few years ago." *Democracy Now!*, "South Korea Cracks Down on Resistance to Jeju Island Naval Base Project," July 19, 2011; "Indian Navy Awaits Regional Nod for Patrolling Malacca Straits," *Hindustan Times*, June 7, 2006. On US-India civilian nuclear deal, see note 5, this chapter.

Chapter 5: China and the Green Revolution

1   Yoni Cohen, "Green Startups Target the Department of Defense," GreentechMedia.com, March 11, 2011; Martin LaMonica, "Five Things We Learned at the ARPA-E Summit," CNet.com, February 29, 2012; and Bruce V. Bigelow, "Navy Draws Heavy Media Coverage for Biggest Biofuel Sea Trial," Xconomy.com, November 21, 2011.

2   Since the late 1980s the Pentagon has been working on transforming the infantry soldier into a complete weapons

system, currently referred to as Future Force Warrior. Work is conducted at MIT (ISN) and UC Berkeley (BLEEX). Based on projected trends in a "future security environment," Future Force Warrior is being readied for climate change and natural disasters, rising resource demands, and the proliferation of WMDs. The program's research investment has "demonstrated commercial spin-off benefits for the nation's civilian economy." US Army Natick Soldier Research, Development & Engineering Center (Nsrdec.Natick.Army.Mil), s.vv. "NSRDEC Future Soldier 2030 Initiative," "Doing Business with Us." On nanotech and federal funding, see notes 11 and 12, chap. 2.

3   The US Weatherization Assistance Program (WAP) is limited to serving low-income households at or below 150 percent of the poverty guidelines. DOE, Energy Efficiency & Renewable Energy, s.v. "Weatherization & Intergovernmental Program."

4   According to Bloomberg, one source of loans and credit lines for China's green technology, CDB, has "more than twice the World Bank's assets." CDB funds Sinovel, Xinjiang Goldwind, Suntech, and China Ming Yang Wind. Natalie Obiko Pearson, "China Targets GE Wind Turbines with $15.5 Billion War Chest," Bloomberg.com, October 14, 2011. The US imported $3.1 billion worth of Chinese solar cells in 2011; in March 2012 the US Department of Commerce announced a tariff on imported Chinese solar cells and panels after seven manufacturers filed a complaint alleging: "illegal government subsidies have made it possible for Chinese companies to gain unfair trade advantages. The subsidies include loans, lines of credit, tax breaks, and favorable terms for insurances, land and utility costs." Ucilia Wang, "Obama Administration to Impose Tariffs on Chinese Solar Panels," *Green Tech* (blog), *Forbes*, March 20, 2012. See also "Green Dragon Fund (GRNDRGN: KY)," Bloomberg.com.

5   See John Tirman, ed., *The Militarization of High Technology* (Cambridge, MA: Ballinger, 1984); Nick Turse, *The Complex: How the Military Invades Our Everyday Lives*, 1st ed. (New York: Metropolitan Books, 2008).

6   The cofounder of Raytheon, Vannevar Bush, joined MIT's Electrical Engineering Department in 1919, eventually serving as dean and vice president. During World War II, he was the chief administrator of the Manhattan Project and served as director of the OSRD, a department he helped initiate during the Roosevelt administration. Bush is credited with codifying the relationship between federally funded science, industry, and the military (i.e., the military-industrial complex). For the blueprint, see *Science, The Endless Frontier: A Report to the President by Vannevar Bush* (Washington, DC: Government Printing Office, 1945). For biographical insight, see Richard Rhodes, *The Making of the Atomic Bomb* (New York: Simon & Schuster, 1988), 336. On Raytheon and ISN, see note 11, chap. 2.

7   India amended its patent law to comply with the WTO Agreement on TRIPS in 2005, though legal battles over medicine patents continue into the present, most notably with Novartis, a Swiss pharmaceutical company with subsidiaries in India. According to Section 3(d) of the Indian Patent Act, "incremental or frivolous innovation is non-patentable." NGOs believe a weakening of Section 3(d) could jeopardize India's capacity to provide affordable generics to the developing world. Rachel Marusak Hermann, "Novartis before India's Supreme Court: What's Really at Stake?," Intellectual Property Watch (IP-Watch.org), March 2, 2012.

8   "The government already spends more than $30 billion a year on bio-medical research through the national institutes of health. It would make much more sense to directly finance the research by the industry, eliminate the tax breaks and let all drugs be sold as generics at Wal-Mart for $4 per prescription." Dean Baker, "Start with the Drug Companies," *Room for Debate* (online forum), *New York Times*, April 13, 2011. See also "Financing Drug Research: What Are the Issues?," Center for Economic and Policy Research (CERP), September 2004.

9   According to Michael J. Graetz, one of the greatest challenges to implementing a successful US energy policy is the

"tendency for Congress to place geographic considerations above technological and economic prospects. . . . Members of Congress frequently have insisted on their own personal priorities, directing funds to individual projects, locations, or institutions by earmarking projects. . . . Clearly, many members of Congress have been more concerned with rewarding well-connected constituents and contributors than advancing science or promising technologies." "Energy Policy: Past or Prologue?," *Daedalus* 141 (Spring 2012), 35.

10   Problems with the United States' "innovation ecosystem" were pointed out by Pres. Bush's scientific advisors in 2004: "Design, product development, and process evolution all benefit from proximity to manufacturing, so that new ideas can be tested and discussed with those working 'on the ground.' . . . The interdependency between new research and manufacturing becomes vitally important, and those linkages are provided by people." President's Council of Advisors on Science and Technology, "Sustaining the Nation's Innovation Ecosystems," January 2004. The Bureau of Labor Statistics projects US employment in manufacturing will continue to trend as an area of rapid decline. Richard Henderson, "Industry Employment and Output Projections to 2020," *Monthly Labor Review*, January 2012.

Chapter 6: Research and Religion (or, The Invisible Hand)

1   Pew's analysis found that among religious groups, the unaffiliated are the most likely to say the earth is warming due to human activities; white evangelical Protestants are the most likely to say there is no solid evidence the earth is warming or that humans play a role; and black Protestants are the least likely to deny global warming is occurring. Another Pew study found views on climate change break along discernible party lines. Pew Research Center (PewResearch.org), "Faith in Global Warming: Religious Groups' Views on Earth Warming Evidence," April 16, 2009; "Wide Partisan Divide over Global Warming: Few Tea Party Republicans See Evidence," October

27, 2010. See also, Aaron M. McCright and Riley E. Dunlap, "The Politicization of Climate Change and Polarization in the American Public's Views of Global Warming, 2001–2010," *Sociological Quarterly* 52 (Spring 2011): 155–94, doi:10.1111/j.1533-8525.2011.01198.x.

2   See profiles of Chesapeake Energy's Aubrey McClendon and Texas billionaire Harold Simmons. Jeff Goodell, "The Big Fracking Bubble: The Scam behind Aubrey McClendon's Gas Boom," *Rolling Stone*, March 15, 2012; Monica Langley, "Texas Billionaire Doles Out Election's Biggest Checks," *Wall Street Journal*, March 23, 2012.

3   During the 2008 presidential campaign John McCain promised to address climate change. By 2011, the majority of Republican presidential candidates denied its existence. Tim Phillips, president of the Koch-funded Americans for Prosperity, chalks it up to the Tea Party and other groups: "If you look at where the situation was three years ago and where it is today, there's been a dramatic turnaround. . . . If you [Republican candidates] . . . buy into green energy or you play footsie on this issue, you do so at your political peril. The vast majority of people who are involved in the nominating process—the conventions and the primaries—are suspect of the science. And that's our influence. Groups like Americans for Prosperity have done it." Coral Davenport, "Heads in the Sand," *National Journal*, December 1, 2011. Americans for Prosperity Director Nansen Malin attended the Saul Alinsky Institute in the early '70s and is writing a book for conservatives on community organizing; she ranks number five on #TCOT ("Top Conservatives on Twitter"). See also FreedomWorks of America's use of Alinsky's *Rules for Radicals*, Brad Knickerbocker, "Who Is Saul Alinsky, and Why Is Newt Gingrich So Obsessed with Him?," *Christian Science Monitor*, January 28, 2012.

4   Max Weber, *The Protestant Ethic and the Spirit of Capitalism*, trans. Talcott Parsons (1904–5; repr., London: Routledge, 1992), 58.

5   In 1834 workers at the Lowell Mills factory, referred to as
     "factory girls," went on strike upon learning wages would be
     reduced by 15 percent. According to the *Boston Transcript*:
     "The number [of strikers] soon increased to nearly 800. A
     procession was formed, and they marched about the town,
     to the amusement of a mob of idlers and boys, . . . We are
     told that one of the leaders mounted a stump and made a
     flaming Mary Wollstonecraft speech on the rights of women
     and the iniquities of the 'monied aristocracy,' which produced
     a powerful effect on her auditors, and they determined to
     'have their way if they died for it.'" Newspaper excerpt from
     historian Catherine Lavender's website, *"Liberty Rhetoric" and
     Nineteenth-Century American Women*, s.v. "Uses of Liberty
     Rhetoric among Lowell Mill Girls." See also title page of the
     workers' paper, the *Lowell Offering*, s.v. "Lives of Lowell
     Mill Girls." For more on the labor press, see *Chomsky on
     MisEducation*, ed. Donaldo Macedo (Lanham, MD: Rowman &
     Littlefield, 2004), chap. 2.

6   See Britain's "Mass Trespass" on Kinder Scout Mountain
     in 1932. The Mass Trespass, led by workers who sought
     unimpeded foot travel, eventually led to the establishment of
     Britain's national parks and the 2004 Right to Roam Act. On
     the history of "rambling" and current campaigns to protect
     walkers' rights, see Ramblers.org.uk.

7   Mitchell Landsberg, "Rick Santorum Denies Questioning
     Obama's Faith," *Los Angeles Times*, February 19, 2012.

8   Richard Land is president of the Southern Baptist Convention's
     Ethics and Religious Liberty Commission and author of *Real
     Homeland Security: The America God Will Bless*. Thomas B.
     Edsall, "Newt Gingrich and the Future of the Right," *Campaign
     Stops* (blog), *New York Times*, January 29, 2012. On Gingrich
     and "Saul Alinsky radicalism," see note 3, this chapter.

9   Smith's *Wealth of Nations* also covers the topic of propagation
     and the health of the labor force: "Poverty, though it does
     not prevent the generation, is extremely unfavourable to the
     rearing of children. The tender plant is produced, but in so

cold a soil, and so severe a climate, soon withers and dies. .
. . In some places one half the children born die before they
are four years of age; in many places before they are seven;
and in almost all places before they are nine or ten. This great
mortality, however, will every where be found chiefly among
the children of the common people, who cannot afford to
tend them with the same care as those of better station."
Smith proposed better wages for workers, enabling families
to better provide for their children, consequently providing a
healthier, more productive workforce. Adam Smith, *An Inquiry
into the Nature and Causes of the Wealth of Nations* (1776; repr.,
London: Methuen, 1904), *Library of Economics and Liberty*
(EconLib.org), s.v. "Adam Smith, Wealth of Nations," s.v. "I.8
Of the Wages of Labour."

Chapter 7: Extraordinary Lives

1   The Russell-Einstein Manifesto, issued in London on July 9,
1955, provided the impetus for the formation of the Pugwash
Conferences which began two years later and continue into the
present. The group derived its name from the location of the
first conference held in Pugwash, Nova Scotia. Membership is
worldwide and follows a basic tenet: "Participation is always
by individuals in their private capacity (not as representatives
of governments or organizations)." Contemporary concerns
include nonproliferation, reduction of chemical and biological
weapons, and the establishment of a NWFZ in the Middle East.
Pugwash Conferences on Science and World Affairs (Pugwash.
org). On Pugwash and Joseph Rotblat, see note 9, this chapter.

2   Barry Feinberg and Ronald Kasrils, *Bertrand Russell's America:
His Transatlantic Travels and Writings: Volume Two, 1945–1970*
(London: George Allen & Unwin, 1984).

3   Lawrence Wittner produced a trilogy of books chronicling the
history of the world nuclear disarmament movement: *One
World or None* (1945–1954), *Resisting the Bomb* (1954–1970),
and *Toward Nuclear Abolition* (1971–present). His most recent
book is *Confronting the Bomb: A Short History of the World*

Nuclear Disarmament Movement (Stanford, CA: Stanford University Press, 2009).

4   Established in London in 1958, the Campaign for Nuclear Disarmament (CND) advocates for Britain's unilateral nuclear disarmament and much more. Early protests took the form of yearly marches to a nuclear weapons facility at Aldermaston. In 1960, some campaign supporters favored sit-ins and blockades, establishing a separate group led by Bertrand Russell, the Committee of 100. (Most Committee of 100 events resulted in arrest.) Contemporary concerns include opposition to the Trident nuclear weapon system, chemical and biological weapons, missile defense, a nuclear-armed NATO, and expansion of nuclear power.

5   Duff's published works include *Prisoners in Vietnam: The Whole Story* (London: ICDP, 1970); *Left, Left, Left: Personal Account of Six Protest Campaigns, 1945–65* (London: Allison & Busby, 1971); *War or Peace in the Middle East* (Nottingham: Spokesman Books, 1978).

6   Carol (née Schatz) Chomsky received her PhD in linguistics from Harvard and served on the faculty of the Harvard Graduate School of Education from 1972 until 1997. She has been described as a "pioneer in the field of child language acquisition," introducing a technique still in use today for helping children learn the mechanics of reading. The technique, referred to as "repeated listening," is discussed in "After Decoding: What?," *Language Arts* 53 (March 1976): 288–96, 314. See also her work on language acquisition by the deaf-blind, *Rich Languages from Poor Inputs,* ed. Massimo Piattelli-Palmarini and Robert C. Berwick (New York: Oxford University Press USA, 2012).

7   "A Call to Resist Illegitimate Authority" appeared in the *New Republic* and the *New York Review of Books* prior to an antidraft demonstration in Washington, DC, in 1967. The organization, Resist, Inc., included members of the clergy and academia who pledged financial support for those choosing to resist the draft, including funds "to supply legal defense and bail." An

archive of the organization's documents, *The Resist Collection*, is held at the Trinity College Library in Hartford, CT.

8   In the 1950s biologist Barry Commoner worked on a project measuring levels of radioactive strontium 90 in the baby teeth of North American children. As a result of incoming data—namely, radioactive fallout from aboveground testing increases radioisotope burden in the biosphere, including bioaccumulation in humans—Commoner and Pauling partnered in writing a petition calling for a ban on nuclear weapons testing in 1957. The petition gained international support, and eventually resulted in the Partial Test Ban Treaty (PTBT). The treaty was not successfully negotiated until 1963, due in part to Edward Teller's insistence on a program of peaceful nuclear explosions (PNEs). See the original petition online at *Linus Pauling and the International Peace Movement*, s.v. "U.S. Signatures to the Appeal by American Scientists to the Governments and People of the World," January 15, 1958. On Edward Teller, see Dan O'Neill, *The Firecracker Boys: H-Bombs, Inupiat Eskimos, and the Roots of the Environmental Movement* (New York: Basic Books, 2007), 296–302. On PNE protest, see Appendix 9.

9   Joseph (Józef) Rotblat was one of two project scientists to leave the Manhattan Project before the bombing of Japan, a move that caused heightened suspicions about his motives. He would spend the remainder of his life calling for the abolition of nuclear weapons and the end to war. After partnering with scientist Yasushi Nishiwaki in deducing the actual fallout from the *Lucky Dragon* incident in 1954, Rotblat worked with Bertrand Russell, playing an instrumental role in the Russell-Einstein Manifesto and the founding of the Pugwash Conferences. On Russell and Committee of 100, see note 4, this chapter.

10   Cf. Steiner-Chomsky exchange, March 23, 1967, and "An Exchange on Resistance: Chad Walsh and William X X, reply by Noam Chomsky," *New York Review of Books*, February 1, 1968.

11 Twenty-five thousand participants attended the "March on Washington to End the War in Vietnam," organized by Students for a Democratic Society (SDS) in 1965. Following several hours of picketing outside the White House, Paul Potter delivered his speech at the Washington Monument. To read Potter's speech, see SDSRebels.com, s.v. "Antiwar Speeches."

12 See King's "Beyond Vietnam." The speech is a hard-hitting analysis of war, militarism, and inequality: "Now there is little left to build on [in Vietnam], save bitterness. Soon, the only solid—solid physical foundations remaining will be found at our military bases and in the concrete of the concentration camps we call 'fortified hamlets.' The peasants may well wonder if we plan to build our new Vietnam on such grounds as these. Could we blame them for such thoughts? We must speak for them and raise the questions they cannot raise. These, too, are our brothers." Martin Luther King Jr., "Beyond Vietnam: A Time to Break Silence," speech at Riverside Church, New York (April 4, 1967).

13 In King's 1968 speech, he called for the development of "a kind of dangerous unselfishness" on behalf of sanitation workers and the building of an allied economic base through boycott, a "bank-in" movement, and an "insurance-in" that encouraged patronage at black-owned businesses. Martin Luther King Jr., "I've Been to the Mountaintop," speech at Bishop Charles Mason Temple, Memphis, TN (April 3, 1968).

14 As part of the "Poor People's Campaign of 1968," Resurrection City was organized, built, and occupied for a span of forty-three days from May to June. An estimated five thousand demonstrators participated in the "live-in" located on the Mall in Washington, DC. The assassinations of Martin Luther King and Robert F. Kennedy took their toll on the campaign, as did torrential rain on the makeshift city. For successes and failures of the action, see John Wiebenson, "Planning and Using Resurrection City," *Journal of the*

American Institute of Planners 35 (November 1969): 405–11, doi:10.1080/01944366908977260.

Chapter 8: MAD (Mutually Assured Dependence)

1   John M. Broder, "Greenpeace Leader Visits Boardroom, without Forsaking Social Activism," *New York Times,* December 7, 2011.

2   A decade prior to the 2010 "People's World Summit on Climate Change and the Rights of Mother Earth," Bolivian activists had successfully resisted an attempt by Aguas del Tunari (a subsidiary of US-based Bechtel) to privatize the water supply. For a detailed account, see Oscar Olivera, *¡Cochabamba! Water War in Bolivia* (Cambridge, MA: South End Press, 2004).

3   Tropical glaciers in the Andean region are at risk, and scientists forecast none will exist in thirty years. Jessica Camille Aguirre, "As Glaciers Melt, Bolivia Fights for the Good Life," *Yes!,* March 18, 2010. See also "Arctic Sea Ice News & Analysis," National Snow & Ice Data Center (NSIDC.org).

4   In August 2012 the US Drought Monitor reported 62.9 percent of the contiguous US as experiencing moderate to exceptional drought, with the percent of the worst categories (extreme to exceptional drought) doubling. As a result of drought conditions, widespread crop failure was reported nationwide, with FAO forecasts of shortages and rising prices worldwide. See also James Hansen et al., "Global Temperature Change," *PNAS* 103, no. 39 (2006): 14288–93, doi:10.1073/pnas.0606291103.

5   See Noam Chomsky, "How the Magna Carta Became a Minor Carta, Part 1 and 2," *Guardian* (London), July 24–25, 2012; "Carte Blanche," TomDispatch.com (audio), July 21, 2012.

6   According to a recent survey by the Yale Project on Climate Change Communication, majorities among six identified groups say climate change and clean energy should be among top national priorities. Yet, according to project director Anthony Leiserowitz, the group with the most influence, climate-change skeptics, account for "only 10 percent [of the

population]" but "appear much larger because they tend to dominate . . . much of the public square." *Talk of the Nation*, "Gauging Public Opinion on Climate Change Policy," NPR, May 4, 2012. On the influence of Koch-funded groups on the election process, see note 3, chap. 6.

7   Shelby Lin Erdman, "Battle over Controversial International Oil Pipeline Growing," CNN, September 5, 2011. The API spokesperson quoted in the article was contacted to verify accuracy; she responded, "If they [Tar Sands Action participants] are protesting the pipeline they are protesting a shovel-ready job that will put thousands of Americans to work. This industry is focused on creating jobs, producing energy responsibly and strengthening America's energy security." Sabrina Fang, API Media Relations, e-mail correspondence, November 16, 2011. On how Saudi interests infuse money into US elections through trade associations, namely, API, see Lee Fang, "How Big Business Is Buying the Election," *The Nation*, September 17, 2012.

8   The Tar Sands Action is part of an ongoing campaign to protest the proposed 1,661-mile pipeline from Alberta, Canada, to refineries on the Texas Gulf Coast. The unconventional product to be conveyed, chemical-laden bitumen derived from the Canadian tar sands, has been described as "the dirtiest oil on the planet." The largest action to date took place in front of the White House between late August and early September 2011. During the two-week sit-in, more than twelve hundred participants committed acts of civil disobedience, resulting in arrest. The event involved a consortium of groups and individuals: Bold Nebraska, Indigenous Environmental Network, 350.org, activists, '08 Obama campaigners, farmers, scientists, and writers.

9   Clifford Krauss, "U.S. Reliance on Oil from Saudi Arabia Is Growing Again," *New York Times*, August 16, 2012. On Saudi plans to refine Canadian tar sands in Texas, see Lee Fang, note 7, this chapter. On history of OPEC, see note 8, chap. 1.

10 Lawrence M. Krauss, "Judgement Day," *New Humanist*, March/
April 2010.

11 During the Geneva Conference in July 1955, Pres. Eisenhower
spoke candidly to representatives from the USSR, telling
Nikolai Bulganin that modern weapons were developed to
the point that any country that used them "genuinely risked
destroying itself. . . . A major war would destroy the Northern
Hemisphere." He made a similar point with Georgi Zhukov:
"Not even scientists could say what would happen if two
hundred H-bombs were exploded in a short period of time,
but . . . the fall-out might destroy entire nations and possibly
the whole northern hemisphere." Francis X. Winters, *The Year
of the Hare: America in Vietnam, January 25, 1963–February 15,
1964* (Athens, GA: University of Georgia Press, 1999), 7–8.

12 Leading up to the 1962 Soviet missile installation, the
Kennedy administration carried out two major covert
operations in Cuba: the Bay of Pigs invasion and Operation
Mongoose. The latter has been described by historian Stephen
G. Rabe as a "massive campaign of terrorism and sabotage."
*The Most Dangerous Area in the World: John F. Kennedy
Confronts Communist Revolution in Latin America* (Chapel Hill:
University of North Carolina Press, 1999), 137. According to
Graham Allison: "The U.S. air strike and invasion that were
scheduled for the third week of the confrontation would likely
have triggered a nuclear response against American ships and
troops, and perhaps even Miami. The resulting war might
have led to the deaths of 100 million Americans and over 100
million Russians." "The Cuban Missile Crisis at 50: Lessons for
U.S. Foreign Policy Today," *Foreign Affairs*, July/August 2012.

13 National Security Archive Electronic Briefing Book No. 281,
s.v. "Documents 8A-D: DEFCON 3 during the October War."

14 The CIA speculates Soviet fears of an imminent attack may
have been a response to US actions launched a few months
into Reagan's first term: air and naval probes near Soviet
borders that sought vulnerabilities in early warning systems;
fleet exercises in proximity to sensitive Soviet military and

industrial sites and operations that simulated surprise naval attacks; radar-jamming and transmission of false radar signals; submarine and antisubmarine aircraft conducting maneuvers in areas where the Soviet Navy stationed its nuclear-powered ballistic missile submarines; and simulated bombing runs over a Soviet military installation in the Kuril Island chain. Central Intelligence Agency (CIA.gov), CSI Publications, s.v. "Books and Monographs," s.v. "A Cold War Conundrum: The 1983 Soviet War Scare," March 19, 2007.

15   In November 2011 Russian Pres. Dmitry Medvedev issued a statement drawing a direct correlation between Pres. Obama's 2009 revision of a missile system—a two-part installation in Poland and the Czech Republic planned by the previous administration—and the willingness of Russia to negotiate the New START treaty. He also stressed any plans for a European missile defense system that excluded Russia from "building a genuine strategic partnership" with NATO could result in withdrawal from START. Medvedev delineated additional measures, and by January 2012, it was reported Iskander missiles had been deployed to Kaliningrad, an exclave between Poland and Lithuania on the Baltic Sea. "Statement in Connection with the Situation concerning the NATO Countries' Missile Defence System in Europe," President of Russia (Kremlin.ru), November 23, 2011; "Russia Starts Deploying Iskander Missiles in Kaliningrad Region," RT (Moscow), January 25, 2012.

16   "Operation Samson: Israel's Deployment of Nuclear Missiles on Subs from Germany," *Der Spiegel*, June 4, 2012.

17   Jerrold Meinwald, "Prelude," *Daedalus* 141 (Summer 2012): 7.

18   Chapter 4, Article 8 of Bolivia's Law No. 071 calls for the promotion of peace and the elimination of all nuclear, chemical, and biological arms and weapons of mass destruction (IV. 8. 6. "Promover la paz y la eliminación de todas las armas nucleares, químicas, biológicas y de destrucción masiva"). For comparison, the Comprehensive Test-Ban Treaty (CTBT), which bans all nuclear explosions, has been signed but will

not enter into force until ten remaining states complete ratification; the US is among the holdouts. On Bolivia's law, see note 1, chap. 1.

Index

Point Hope letter of, 123, 124–26,
127–28
by scientists for crops destroyed in
Vietnam, 99–102
Tar Sands Action, 84–85, 158n7,
159n8
Pugwash Conferences, 154n1, 156n9

al-Qaeda, 24, 25, 136n3

Rabe, Stephen G., 160n12
racism, nuclear, 48–49
radar, 139n13, 160n14
Radiation Laboratory (RadLab),
139n13
radioactivity, 90–94, 103–6, 142n14
RadLab. See Radiation Laboratory
Rashid, Ahmed, 136n3
Raytheon, 30, 55, 58, 138n11, 149n6
R&D (research & development), 60,
138n11, 150n8
Rea (Lt. Col.), 89–95
Reagan, Ronald, 22, 42, 68, 86,
136n2, 160n14
    with Islamic militants and Zia alli-
    ance, 136n3
    with massacre cover ups, 23
    with radical Islam, 24
    "Star Wars" program and, 33, 74
Real Homeland Security: The America
God Will Bless (Land), 153n8
religion. See also specific religions
    with Christians and religious right,
    63–65, 67–68, 151n1
    research and, 63–71
religious right, 63–64, 65, 67–68,
134n6, 151n1, 158n6
Republicans, climate change and, 15,
17–18, 152n3
research
    with humans and radioactivity,
    103–6, 142n14
    manufacturing linked to, 151n10
    religion and, 63–71
    toxicity of war and politics influ-
    encing scientific, 34–35
research & development. See R&D
Research Laboratory of Electronics
(RLE), 139n13

"The Responsibility of Intellectuals"
(Chomsky), 76–77
Resurrection City, 79, 157n14
Ricardo, David, 70–71
rights, 14, 78–79, 82, 83, 152n5,
157n2
Right to Roam Act of 2004, 153n6
RLE. See Research Laboratory of
Electronics
Roosevelt, Franklin D., 49–50,
146n15, 149n6
Rotblat, Joseph, 156n9
Rumsfeld, Donald, 21
Russell, Bertrand, 73–74, 76, 154n4,
156n9
Russell-Einstein Manifesto, 74,
154n1, 156n9
Russia, 41, 50, 86, 87, 137n4, 143n1,
161n15
ruthlessness, in acquisition, 66–67

sanctions, 50–51
Santorum, Rick, 67
Saudi Arabia, 16, 24, 25, 134n8,
136n3
science
    economics over, 116
    future possibilities in, 87, 138n12
    with lobbyists swaying public
    opinion about climate change,
    18–19, 135n11
    nanoscale, 138n11
scientists, 76, 99–102
Scorched Earth: Legacies of Chemical
Warfare in Vietnam (Wilcox), 34
SDS. See Students for a Democratic
Society
sea. See deep-sea drilling
secrecy
    with atomic-bomb deaths, 89–95
    corporate funding and imposed,
    28
    with Israel's nuclear ambiguity,
    145n9
    with radioactive releases, 38
    with US and NSG, 144n5
Simpson, Alan, 23
Singh, Manmohan, 143n2
Smith, Adam, 70, 83, 134n6, 153n9

About the Authors

NOAM CHOMSKY was born in Philadelphia in 1928. He studied at the University of Pennsylvania where he received his PhD in linguistics in 1955. He joined the staff at MIT and was appointed Institute Professor in 1976, gaining international renown for his theories on the acquisition and generation of language. He became well known as an activist and public intellectual during the Vietnam War; he became known as a formidable critic of media with the 1988 release of *Manufacturing Consent*, a book coauthored with Edward Herman. With the publication of *9/11* in November 2001, inarguably one of the most significant books on the subject, he became as widely read and as an essential a voice internationally as other political philosophers throughout history. That book, like the present volume, was composed from interviews. Chomsky has written and lectured widely on linguistics, philosophy, intellectual history, contemporary issues, international affairs, and US foreign policy. In 2010 Chomsky, Eduardo Galeano, Michael Hardt, Naomi Klein, and Vandana Shiva became signatories to United for Global Democracy, a manifesto created by the international Occupy movement.

LARAY POLK was born in Oklahoma in 1961 and currently lives in Dallas, Texas. She is a multimedia artist and writer. Her articles and investigative reports have appeared in the *Dallas Morning News*, *D Magazine*, and *In These Times*. As a 2009 grant recipient from the Investigative Fund at the Nation Institute, she produced stories on the political entanglements and compromised science behind the establishment of a radioactive waste disposal site in Texas, situated in close proximity to the Ogallala Aquifer.

About Seven Stories Press

Seven Stories Press is an independent book publisher based in New York City. We publish works of the imagination by such writers as Nelson Algren, Russell Banks, Octavia E. Butler, Ani DiFranco, Assia Djebar, Ariel Dorfman, Coco Fusco, Barry Gifford, Martha Long, Luis Negrón, Hwang Sok-yong, Lee Stringer, and Kurt Vonnegut, to name a few, together with political titles by voices of conscience, including Subhankar Banerjee, the Boston Women's Health Collective, Noam Chomsky, Angela Y. Davis, Human Rights Watch, Derrick Jensen, Ralph Nader, Loretta Napoleoni, Gary Null, Greg Palast, Project Censored, Barbara Seaman, Alice Walker, Gary Webb, and Howard Zinn, among many others. Seven Stories Press believes publishers have a special responsibility to defend free speech and human rights, and to celebrate the gifts of the human imagination, wherever we can. In 2012 we launched Triangle Square *books for young readers* with strong social justice and narrative components, telling personal stories of courage and commitment. For additional information, visit www.sevenstories.com.